# ROBOT SYSTEM RELIABILITY AND SAFETY

# ROBOT SYSTEM RELIABILITY AND SAFETY

## A MODERN APPROACH

B.S. DHILLON

CRC Press
Taylor & Francis Group
Boca Raton London New York

CRC Press is an imprint of the
Taylor & Francis Group, an **informa** business

CRC Press
Taylor & Francis Group
6000 Broken Sound Parkway NW, Suite 300
Boca Raton, FL 33487-2742



Version Date: 20150318

International Standard Book Number-13: 978-1-4987-0644-5 (Hardback)



**Visit the Taylor & Francis Web site at**
**http://www.taylorandfrancis.com**

**and the CRC Press Web site at**
**http://www.crcpress.com**

*This book is affectionately dedicated to my grandson, Dillon.*

# Contents

# Preface

Today, a vast number of robots are being used to perform various types of tasks around the world. In the industrial sector, robots are often used to perform tasks such as spot-welding, materials-handling, and arc welding. A robot has to be safe and reliable. An unreliable robot may become the cause of unsafe conditions, inconvenience, high maintenance cost, and so on. As robots make use of electrical, electronic, mechanical, hydraulic, and pneumatic parts, this makes the issue of their reliability very challenging because of the many different sources for potential failures.

Needless to say, robot-system reliability and safety have become an important issue. For example, with regard to robot-system safety, a study reported that 12%–17% of the accidents in industries using advanced manufacturing technology were concerned with automated production equipment.

Although, over the years, a large number of articles have appeared in journals and conference proceedings on robot reliability, safety and associated areas, to the best of author's knowledge, since 1991, there has been no book that covers all these topics within its framework. This causes a great deal of difficulty for up-to-date information-seekers because they have to consult many different and diverse sources.

Thus, the main objective of this book is to combine all these topics into a single volume and to eliminate the need to consult many different and diverse sources in obtaining up-to-date desired information on the topics. The sources of most of the material presented are given in the reference section at the end of each chapter. This will be useful to readers if they desire to delve more deeply into a specific area or topic. The book contains a chapter on mathematical concepts and another chapter on the basics of reliability and safety considered useful to understand the content of subsequent chapters. Furthermore, another chapter is devoted to methods considered useful to analyze the reliability and safety of robot systems.

The topics covered in the book are treated in such a manner that the reader will require no previous knowledge to understand the contents. At appropriate places, the book contains examples along with their solutions, and there are numerous problems at the end of each chapter to test the reader's comprehension in the area. An extensive list of publications dating 1974–2013, directly or indirectly on robot-system reliability and safety, is provided at the end of this book to give readers a view of the intensity of developments in the area.

The book comprises 11 chapters. Chapter 1 presents the need for and historical developments in robot-system reliability and safety; robot-system reliability/safety-related facts, figures, and examples; important terms and definitions; and useful sources for obtaining information on robot-system reliability and safety. Chapter 2 reviews mathematical concepts considered useful to understand subsequent chapters. Some of the topics covered in the chapter are Boolean algebra laws, probability properties, probability distributions, and useful definitions.

Chapter 3 presents various introductory aspects of reliability and safety. Chapter 4 presents a number of methods considered useful to analyze robot-system reliability and safety—failure modes and effect analysis, the Markov method, fault tree analysis, technique of operations review, hazard and operability analysis, interface safety analysis, and the probability-tree method. Chapter 5 is devoted to robot-reliability. Some of the topics covered in the chapter are classifications of robot failures and their causes, robot-reliability measures, robot-reliability analysis methods, and hydraulic and electric robots' reliability analysis.

Chapters 6 and 7 are devoted to robot-related safety and accidents and analysis of these. Chapter 6 covers topics such as robot-related safety problems and hazards, roles of robot manufacturers and users in robot-related safety, safety considerations in robot design, installation, programming, and operation and maintenance phases; robot safeguard approaches, common robot safety features and their functions, and robot welding operations-related safety considerations. Some of the topics covered in Chapter 7 are examples of robot-related accidents, accident causes and sources, useful recommendations to prevent human injury by robots, and methods for performing robot-related accident analysis.

Chapter 8 presents various important aspects of robot-maintenance and areas of robotics applications in maintenance and repair, including robot-maintenance-related needs and maintenance types, diagnosis and monitoring approaches, useful guidelines to safeguard robot-maintenance personnel, and models for performing robot-maintenance analysis. Chapter 9 is devoted to human factors and safety considerations in robotics workplaces and covers topics such as human factors-related issues during the robotic systems' factory-integration process, common robot and human tasks, rules of robotics with regard to humans, humans at risk from robots, useful guidelines to safeguard robot operators and teachers, methods to limit robot movements, and methods for performing safety and human error analysis in robotics workplaces.

Chapter 10 is devoted to testing robots, costing, and failure data. Some of the topics covered in the chapter are testing for performance of robots, testing methods, robot-reliability testing, project cost, models for estimating robot-related costs,, failure-data, failure-reporting and documentation system for robots, main data sources for reliability, and data for mean time between robot failures. Finally, Chapter 11 presents six mathematical models for analyzing robot-system reliability and safety.

The book will be useful to many individuals, including system engineers, design engineers, reliability engineers, safety engineers, engineering professionals working in the area of robotics, researchers and instructors involved with robotic systems, graduate and senior undergraduate students in system engineering, reliability engineering, and safety engineering, and engineers-at-large.

The author is deeply indebted to many individuals, including family members, colleagues, friends, and students for their invisible input. The unseen contributions of my children also are appreciated. Last, but not the least, I thank my wife, Rosy, my other half and friend, for typing this entire book and for her timely help in proofreading.

**B. S. Dhillon**
*Ottawa, Ontario, Canada*

# Author

**Dr. B. S. Dhillon** is a professor of engineering management in the Department of Mechanical Engineering at the University of Ottawa, Ontario, Canada. He has served as chairman/director of the Mechanical Engineering Department/Engineering Management Program for over 10 years at the same institution. He is the founder of the probability distribution model named the *Dhillon Distribution/Law/Model* by statistical researchers around the world. He has published over 376 articles (i.e., 226 [70 single-authored + 156 coauthored] journals and 150 conference proceedings) on reliability engineering, maintainability, safety, engineering management, and so on. He is or has been on the editorial boards of 12 international scientific journals. In addition, Dr. Dhillon has written 42 books on various aspects of healthcare, engineering management, design, reliability, safety, and quality, published by Wiley (1981), Van Nostrand (1982), Butterworth (1983), Marcel Dekker (1984), Pergamon (1986), and so on. His books are being used in over 100 countries and many of them are translated into languages such as German, Russian, Chinese, and Persian (Iranian).

He has served as general chairman of two international conferences on reliability and quality control held in Los Angeles and Paris in 1987. Dr. Dhillon has also served as a consultant in various organizations and bodies and has many years of experience in the industrial sector. At the University of Ottawa, he has been teaching reliability, quality, engineering management, design, and related areas for over 35 years and he has also lectured in over 50 countries, having delivered keynote addresses at various international scientific conferences held in North America, Europe, Asia, and Africa. In March 2004, Dr. Dhillon was a distinguished speaker at the Conference/Workshop on Surgical Errors (sponsored by the White House Health and Safety Committee and Pentagon), held on Capitol Hill (One Constitution Avenue, Washington, DC).

Dr. Dhillon attended the University of Wales, Cardiff, Wales, U.K., where he received a BS in electrical and electronic engineering and an MS in mechanical engineering. He received a PhD in industrial engineering from the University of Windsor, Ontario, Canada.

# 1

# *Introduction*

## 1.1 Background

Today, millions of robots are being used around the globe to perform various types of tasks, including materials-handling, spot-welding, arc-welding, and routing. A robot may simply be described as a mechanism guided by automatic controls. The idea of the functional robot may be traced back to the writings of Aristotle (4th century BC), the teacher of Alexander the Great, in which he wrote "If every instrument could accomplish its own work, obeying or anticipating the will of others …." [1].

The word "robot" is derived from the Czechoslovakian language, in which it means "worker" [2]. The first commercial robot was manufactured by Planet Corporation, in 1959 [3]. As robots use mechanical, pneumatic, hydraulic, electrical, and electronic parts, their reliability-related problems are very challenging because of the sources of failures are many and varied. Although there is no clear-cut definitive point for the beginnings of the field of robot reliability, articles by J.F. Engelberger and K.M. Haugan, in 1974, could be regarded as its starting point [4,5]. In 1987, an article presented a comprehensive list of publications that were directly or indirectly related to robot reliability [6] and, in 1991, a book titled *Robot Reliability and Safety* covered the topic of robot reliability in considerable depth [7].

Many articles, directly or indirectly, in the 1970s and early 1980s covered the topic of robot safety [6]; in 1985 and 1986, two standards titled "An Interpretation of the Technical Guidance on Safety Standards in the Use, etc. of Industrial Robots" and "Industrial Robots and Robot Systems—Safety Requirements" were published by the Japanese Industrial Safety and Health Association [8], and the American National Standards Institute [9], respectively. Also, in 1985, an edited book titled *Robot Safety* was published [10].

Needless to say, since the 1980s, a large number of publications have appeared that are directly or indirectly related to robot-system reliability and safety. A list of over 370 such publications is provided in the Appendix.

## 1.2 Robot System Reliability/Safety-Related Facts, Figures, and Examples

Some of the facts, figures, and examples directly or indirectly concerned with robot-system reliability/safety are as follows:

- As per [11], some studies conducted in Japan indicate that more than 50% of working accidents with robots can be attributed to faults in the control system's electronic circuits.
- As per [11], the Japanese studies also indicate that "human error" was responsible for less than 20% of working accidents with robots.
- A study reported that 12–17% of the accidents in industries using advanced manufacturing technology were related to automated production equipment [12,13].
- The first robot-related fatal accidents occurred in Japan in 1978 and in the United States in 1984 [14,15].
- A material-handling robot was operating in its automatic mode and a worker violated all safety devices in entering the robot work-cell. The worker got trapped between the robot and a post anchored to the floor, was consequently injured and died a few days later [7,16,17].
- For the period 1978–1987, there were 10 robot-related fatal accidents in Japan [15].
- A worker entered the robot cell to clean its sensors and was killed because he/she ignored did not observe stated lockout procedures [7,16–20].
- As per [12], during the period 1978–1984, there were at least five robot-related fatal accidents: one in the United States and four in Japan.
- A maintenance worker climbed over a safety fence without turning off robot power and performed tasks in the robot work area while it was temporarily stopped. When the robot recommenced operation, it pushed the maintenance worker into a grinding machine and the worker died subsequently [7,16,17].
- In 1987, a study of 32 robot-related accidents that occurred in four countries—Japan, United States, West Germany, and Sweden—reported that line workers were at the greatest risk of injury followed by maintenance workers [14].
- A worker stepped between a robot and the machine (a planer) it was servicing, and turned off the power to the circuit transmitting/

activating signal from the planer to the robot. After accomplishing the necessary task, the worker switched on the power to the same circuit while still in the robot's work area. When the robot recommenced its operation, it crushed the worker to death against the planer [7,16,17].

- A robot operator went to check/fix a robot failure without locking out the robot. During the process, the operator activated the robot and the arm of the robot crushed the operator against a part being transported on a conveyor and the operator died [7,16–20].
- A worker holding the robot controls in his/her hand activated the robot while bending over the wheel to check the settings. The robot pinned the worker against the wheel and crushed him/her to death [7,16–20].
- A worker turned on a welder-robot while another worker was still in the robot's work area. The robot pushed the worker in its work area into the positioning fixture to death [7,16–20].
- A worker climbed onto a conveyor belt in motion for recovering a faulty part when the robot servicing the line was stopped on a program point temporarily. The robot crushed the worker to death when the belt resumed its operation [7,16–20].
- A worker entered the caged robotic palletizer cell while the robotic palletizer was still running. The worker's torso was crushed by the arms of the robotic palletizer as it tried to pick up boxes on the roller conveyor and the worker died [7,16–20].
- A worker was freeing a jam at the end of an oven and became caught between the conveyor belt of the oven and the robotic arm and got killed [7,16–20].
- A worker was troubleshooting a robotic arm used for removing CD jewel cases from an injection molding machine, when the arm cycled and struck the worker, who died two weeks later [7,16–20].

## 1.3 Terms and Definitions

This section presents some useful terms and definitions directly or indirectly concerned with the reliability and safety of robot systems [7,9,21–29].

- *Robot*. This is an automatic, reprogrammable, position-controlled, multifunctional manipulator, consisting of several axes designed for moving materials, components, tools, or specialized devices through variable programmed motions to perform various tasks.

- *Reliability.* This is the probability that an item/system will perform its stated mission satisfactorily for the specified time period when used according to the specified conditions.
- *Safety.* This is conservation of human life and the prevention of damage to items as per mission-stated requirements.
- *End-effector.* This is a gripper, actuator, or driven mechanical device designed for attachment to the robot wrist (i.e., the end of a manipulator) by which items can be acted upon.
- *Pendant.* This is a portable control device, including teaching pendants, that permits a human to control the robot from within its (i.e., the robot's) work area or zone.
- *Barrier.* This is a physical means of separating humans from the prohibited robot work zone or area.
- *Gripper.* This is the grasping hand of the robot which manipulates items and tools to accomplish a specified task.
- *Error recovery.* This is the capability of intelligent robotic systems to reveal various types of errors and, through programming, to undertake corrective actions to overcome the impending problem and complete the specified process.
- *Erratic robot.* A robot that has moved appreciably off its stated path.
- *Hazardous motion.* This is a sudden robot motion that may result in injury.
- *Fail-safe.* This is the failure of a robot/robot part, without endangering humans or damage to equipment or plant facilities.
- *Robot mean time to failure.* This is the average time that a robot will operate before failure.
- *Awareness barrier.* This is a device or an attachment that, by physical and visual means, alerts humans regarding potential or existing hazards.
- *Robot mean time to repair.* This is the average time that a robot is expected to be out of action after failure.
- *Safeguard.* This is a barrier device, guard, or procedure developed for human protection.
- *Downtime.* This is the time period during which the item/system/robot is not in a condition to carry out its stated mission.
- *Mission time.* This is the element of uptime that is required to perform a stated mission profile.
- *Random failure.* This is any failure whose occurrence cannot be predicted.
- *Redundancy.* The existence of more than one means to perform a specified function.

- *Accident*. This is an event that involves damage to a certain system/item/robot that suddenly disrupts the potential or current system/item/robot output.
- *Continuous task*. This is a task/job that involves some kind of tracking activity (i.e., monitoring a changing condition).

## 1.4 Useful Sources for Obtaining Information on Reliability and Safety of Robot Systems

This section lists journals, conference proceedings, books, technical reports, data sources, standards, and organizations directly or indirectly considered quite useful to obtain information on the reliability and safety of robot systems.

### 1.4.1 Journals

- *Robotica*
- *Robotics*
- *Journal of Occupational Accidents*
- *Safety Science*
- *Accident Analysis and Prevention*
- *International Journal of Human Factors Engineering*
- *Journal of Robotic Systems*
- *Robotics World*
- *The Industrial Robot*
- *Soviet Engineering Research*
- *International Journal of Robotics and Automation*
- *Journal of Safety Research*
- *Journal of Mechanical Design*
- *Professional Safety*
- *Reliability Engineering and System Safety*
- *IEEE Transactions on Reliability*
- *Robotics Engineering*
- *Microelectronics and Reliability*
- *Robotics and Autonomous Systems*
- *International Journal of Human Factors in Manufacturing*

- *Robotics Today*
- *International Journal of Robotics Research*
- *IEEE Transactions on Robotics and Automation*

### 1.4.2 Conference Proceedings

- *Proceedings of the IEEE International Conferences on Robotics and Automation*
- *Proceedings of the Annual Meetings of the Human Factors and Ergonomics Society*
- *Proceedings of the British Robot Association Annual Conferences*
- *Proceedings of the International Conferences on CAD/CAM, Robotics and Factories of the Future*
- *Proceedings of the IEEE International Conferences on Systems, Man, and Cybernetics*
- *Proceedings of the Annual Reliability and Maintainability Symposium*
- *Proceedings of the ANS Topical Meetings on Robotics and Remote Systems*
- *Proceedings of the IEEE International Symposium on Robot and Human Interactive Communication*

### 1.4.3 Books

- Bonney, M.C., Yong, Y.E., Editors, *Robot Safety*, Springer-Verlag, New York, 1985.
- Graham, J.H., Editor, *Safety, Reliability, and Human Factors*, Van Nostrand Reinhold, New York, 1991.
- Dhillon, B.S., *Robot Reliability and Safety*, Springer-Verlag, New York, 1991.
- Nof, S.Y., Editor, *Handbook of Industrial Robotics*, John Wiley and Sons, New York, 2007.
- Kurfess, T.R., Editor, *Robotics and Automation Handbook*, CRC Press, Boca Raton, Florida, 2005.
- Siciliano, B., Khatib, O., Editor, *Handbook of Robotics*, Springer-Verlag, Berlin, 2008.
- Dhillon, B.S., *Design Reliability: Fundamentals and Applications*, CRC Press, Boca Raton, Florida, 1999.
- Dhillon, B.S., *Engineering Safety: Fundamentals, Techniques, and Applications*, World Scientific Publishing, River Edge, New Jersey, 2003.
- Tver, D.F., Bolz, R.W., *Robotics Sourcebook and Dictionary*, Industrial Press, New York, 1983.

- Noro, K., Editor, *Occupational Health and Safety in Automation and Robotics*, Taylor & Francis, London, 1987.
- Shooman, M.L., *Probabilistic Reliability: An Engineering Approach*, McGraw Hill Book Company, New York, 1968.
- Handley, W., *Industrial Safety Handbook*, McGraw Hill Book Company, New York, 1969.
- Geotsch, D.L., *Occupational Safety and Health*, Prentice-Hall, Englewood Cliffs, New Jersey, 1996.
- Stephans, R.A., Talso, W.W., Editors, *System Safety Analysis Handbook*, System Safety Society, Irvine, California, 1993.
- Dhillon, B.S., *Reliability, Quality, and Safety for Engineers*, CRC Press, Boca Raton, Florida, 2005.
- Siegwart, R., Nourbakhsh, I.R., *Introduction to Autonomous Mobile Robots*, MIT Press, Cambridge, Massachusetts, 2011.


### 1.4.4 Technical Reports


- Hetzler, W.E., Hirsh, G.L., Machine operator crushed by robotic platform, *Nebraska Fatality Assessment and Control Evaluation (FACE) Investigation Report No. 99NE017*, The Nebraska Department of Labor, Omaha, Nebraska, October 25, 1999.
- Ulrich, K.T., Tuttle, T.T., Donoghue, J.P., Townsend, W.T., Intrinsically safer robots, *NASA Report No. NAS 10-12178*, Bararett Technology, Inc., Cambridge, Massachusetts, May 1995.
- Addison, J.H., Robotic safety systems and methods: Savannah River Site, *Report No. DPST-84-907 (DE 35-008261)*, December 1984, issued by E.E. du Pont de Nemonts and Co., Savannah River Laboratory, Aiken, South Carolina.
- National Institute of Occupational Safety and Health (NIOSH), Preventing the injury of workers by robots, *DHHS (NIOSH) Publication No. 85-103*, Morgantown, West Virginia, 1984.
- Japanese Ministry of Labor, Study on accidents involving robots, *Report No. PB 83239822*, Tokyo, 1982, available from the National Technical Information Service (NTIS), Springfield, Virginia.


### 1.4.5 Data Sources

- Reliability Analysis Center, Rome Air Development Center (RADC), Griffis Air Force Base, Rome, New York.
- Government/Industry Data Exchange Program (GIDEP), GIDEP Operations Center, US Department of the Navy, Corona, California.

- International Occupational Safety and Health Information Center Bureau, International du Travail, CH-1211, Geneva 22, Switzerland.
- Safety Research Information Service (SRIS), National Safety Council, 444 North Michigan Avenue, Chicago, Illinois.
- MIL-HANDBOOK 217, Rome Air Development Center, Griffis Air Force Base, Department of Defense, Rome, New York.
- Computer Accident/Incident Report System, System Safety Development Center, EG&G, PO Box 1625, Idaho Falls, Ohio.
- National Technical Information Service (NTIS), United States Department of Commerce, 5285 Port Royal Road, Springfield, Virginia.
- American National Standards Institute (ANSI), 11 W. 42nd St., New York, 10036.

### 1.4.6 Standards

- ANSI/RIA R15.06-2012, *American National Standard for Industrial Robots and Robot Systems-Safety Requirements*, American National Standards Institute, New York, 2012.
- *An Interpretation of the Technical Guidance on Safety Standards in the Use, etc., of Industrial Robots*, Japanese Industrial Safety and Health Association, Tokyo, 1985.
- RIA TR T15.106-2006, *Technical Report for Industrial Robots and Robot Systems-Safety Requirements Teaching Multiple Robots*, Robotic Industries Association, Ann Arbor, Michigan, 2006.
- MIL-STD-721, *Definitions of Terms for Reliability and Maintainability*, Department of Defense, Washington, DC.
- MIL-STD-882, *Systems Safety Program for System and Associated Subsystem and Equipment-Requirements*, Department of Defense, Washington, DC.
- MIL-STD-785, *Reliability Program for Systems and Equipment, Development and Production*, Department of Defense, Washington, DC.
- IEC 60950, *Safety of Information Technology Equipment*, International Electro-Technical Commission, Geneva, Switzerland, 1999.
- ISO 13482: 2014 (en), *Robots and Robotic Devices-Safety Requirements for Personal Care Robots*, International Organization for Standardization (ISO), Geneva 20, Switzerland, 2014.
- CAN/CAS-Z434-03 (R2013), *Industrial Robots and Robot Systems—General Safety Requirements*, Canadian Standards Association, Toronto, Canada, 2013.
- MIL-STD-2155, *Failure Reporting, Analysis, and Corrective Action (FRACAS)*, Department of Defense, Washington, DC.

- MIL-STD-756, *Reliability Modeling and Prediction*, Department of Defense, Washington, DC.
- MIL-STD-1629, *Procedures for Performing Failure Mode, Effects and Criticality Analysis*, Department of Defense, Washington, DC.

### 1.4.7 Organizations

- American Society of Safety Engineers, 1800 East Oakton St., Des Plaines, Illinois.
- British Safety Council, 62 Chancellors Road, London, U.K.
- Reliability Society, IEEE, PO Box 1331, Piscataway, New Jersey.
- Occupational Safety and Health Administration, U.S. Department of Labor, 200 Constitution Avenue, Washington, DC.
- The Robotics Society of America, PO Box 1205, Danville, California 94 526-1205.
- The International Federation of Robotics, Lyoner Str. 18, Frankfurt, Germany.
- Advanced Robot Technology Research Association, Kikai-shinko Bldg, 3-5-8 Shiba-Kohen, Minato-ku, Tokyo, Japan.
- American Society of Mechanical Engineers, 345 E. 47th Street, New York, New York 10017.
- Institute of Electrical and Electronics Engineers, Service Center, 445 Hoes Lane, Piscataway, New Jersey 08854-4150.
- Robotic Industries Association, 900 Victors Way, PO Box 3724, Ann Arbor, Michigan 48106.
- British Robot Association, BRA Aston Science Park, Love Lane, Birmingham, U.K.
- Japan Industrial Robot Association, 3-5-8 Shibakoen, Minato-Ku, Tokyo, Japan.

## 1.5 Scope of the Book

Robots are increasingly being used around the globe to perform various types of tasks. Some examples of such tasks are materials-handling, spot-welding, routing, and arc-welding. A robot has to be safe and reliable; today, reliability and safety of robot systems have become pressing issues because of the increasing number of accidental deaths and injuries; and cost.

Over the years, a large number of publications (i.e., journal and conference proceedings articles, technical reports, etc.) on robot system reliability and safety have appeared in the literature. At present, to the best of the author's knowledge, there is no book on the topic that covers recent developments in the area. Therefore, this book not only attempts to cover the reliability and safety of robot system within its framework, but also provides the latest developments in the area.

Finally, the main objective of this book is to provide professionals and others concerned with robot systems up-to-date information that could be useful for improving reliability and safety of such systems. This book will be useful to many individuals, including professionals working on reliability and safety in the robotics-related industrial sector, researchers, instructors, and graduate and senior undergraduate students in the area of robotics, and engineers-at-large.

## 1.6 Problems

1. Write an essay on reliability and safety of robot systems.
2. Define the following four terms:
   - Robot
   - End-effector
   - Safety
   - Reliability
3. List five important facts and figures concerning the reliability and safety of robot systems.
4. List at least six useful data sources for obtaining information related to the reliability and safety of robot systems.
5. What is the difference between the terms "awareness barrier" and "safeguard?"
6. List at least five books considered important to obtain information related to the reliability and safety of robot systems.
7. Define the following terms:
   - Robot mean time to failure
   - Accident
   - Hazardous motion
   - Error recovery
8. List at least six important organizations to obtain information concerning the reliability and safety of robot systems.
9. Discuss at least five important examples of robot-related accidents.
10. List at least six important standards directly or indirectly concerned with the reliability and safety of robot systems.

## References


1. Heer, E., Robots in modern industry, in *Recent Advances in Robotics*, edited by G. Beni and S. Hackwood, John Wiley and Sons, New York, 1985, pp. 11–36.
2. Jablonowski, J., Posey, J.W., Robotics terminology, in *Handbook of Industrial Robotics*, edited by S.Y. Nof, John Wiley and Sons, New York, 1985, pp. 1271–1303.
3. Zeldman, M.I., *What Every Engineer Should Know About Robots*, Marcel Dekker, New York, 1984.
4. Engelberger, J.F., Three million hours of robot field experience, *Industrial Robot*, Vol. 1, No. 4, 1974, pp. 164–168.
5. Haugan, K.M., Reliability in industrial robots for spray gun applications, *Proceedings of 2nd Conference on Industrial Robot Technology*, University of Birmingham, U.K., 1974, pp. 1–5.
6. Dhillon, B.S., On robot reliability and safety: Bibliography, *Microelectronics and Reliability*, Vol. 27, No. 1, 1987, pp. 105–118.
7. Dhillon, B.S., *Robot Reliability and Safety*, Springer-Verlag, New York, 1991.
8. *An Interpretation of the Technical Guidance on Safety Standards in the Use, etc., of Industrial Robots*, Japanese Industrial Safety and Health Association, Tokyo, 1985.
9. ANSI/RIA R15.06-1986, *American National Standard for Industrial Robots and Robot Systems–Safety Requirements*, American National Standards Institute, New York, 1986.
10. Bonney, M.C., Yong, Y.F., Editors, *Robot Safety*, Springer-Verlag, New York, and IFS (Publications), Bedford, England, 1985.
11. Retsch, T., Schmitter, G., Marty, A., Safety principles for industrial robots, in *Encyclopedia of Occupational Health and Safety*, Vol. II, edited by J.M. Stellman, International Labor Organization, Geneva, Switzerland, 2011, pp. 58.56–58.58.
12. Backtrom, T., Dooes, M., A comparative study of occupational accidents in industries with advanced manufacturing technology, *International Journal of Human Factors in Manufacturing*, Vol. 5, 1995, pp. 267–282.
13. Clark, D.R., Lehto, M.R., Reliability, maintenance, and safety of robots, in *Handbook of Industrial Robotics*, edited by S.Y. Nof, John Wiley and Sons, New York, 1999, pp. 717–753.
14. Sanderson, L.M., Collins, J.N., McGlothlin, J.D., Robot-related fatality involving a U.S manufacturing plant employee: Case report and recommendations, *Journal of Occupational Accidents*, Vol. 8, 1986, pp. 13–23.
15. Nagamachi, M., Ten fatal accidents de to robots in Japan, in *Ergonomics of Hybrid Automated Systems*, edited by W. Karwowski et al., Elsevier, Amsterdam, 1988, pp. 391–396.
16. Japanese Ministry of Labor, *Study on Accidents Involving Industrial Robots, Report No. PB 83239822*, Tokyo, 1982, available from the National Technical Information Service (NTIS), Springfield, Virginia.
17. Nicolaisen, P., Safety problems related to robots, *Robotics*, Vol. 3, 1987, pp. 205–211.
18. Altamuro, V.M., Working safely with the iron collar worker, *National Safety News*, July 1983, pp. 38–40.
19. Occupational Safety and Health Administration (OSHA), *Report No. 0552652*, Washington, DC, October 10, 2006.

20. Lauch, K.E., New Standards for Industrial Robot Safety, *CIM Review*, Spring 1986, pp. 60–68.
21. Fisher, E.L., ed., Glossary of robotics terminology, in *Robotics*, Industrial Engineering and Management Press, Institute of Industrial Engineers, Atlanta, Georgia, 1983, pp. 231–253.
22. Tver, D.F., Boltz, R.W., *Robotics Source Book and Dictionary*, Industrial Press, New York, 1983.
23. Susnjara, K.A., *A Manager's Guide to Industrial Robots*, Corinthian Press, Shaker Heights, Ohio, 1982.
24. US Department of Defense, *MIL-STD-721B, Definitions of Effectiveness Terms for Reliability, Maintainability, Human Factors, and Safety*, Washington, DC, August 1966.
25. American Society of Safety Engineers, *Dictionary of Terms Used in the Safety Profession*, 3rd Edition, Des Plaines, Illinois, 1988.
26. McKenna, T., Oliverson, R., *Glossary of Reliability and Maintenance Terms*, Gulf Publishing Company, Houston, Texas, 1997.
27. Meulen, M.V.D., *Definitions for Hardware and Software Safety Engineers*, Springer-Verlag, London, 2000.
28. Dhillon, B.S., *Engineering Safety: Fundamentals, Techniques, and Applications*, World Scientific Publishing, River Edge, New Jersey, 2003.
29. Dhillon, B.S., *Applied Reliability and Quality: Fundamentals, Methods, and Procedures*, Springer-Verlag, London, 2007.

# 2

# Basic Mathematical Concepts

## 2.1 Introduction

As in the development of other areas of science and technology, mathematics has played an important role in the development of the reliability and safety fields of the robot systems also. The history of mathematics may be traced back to the development of our currently used number symbols, sometimes referred as the "Hindu-Arabic numeral system" in the published literature [1]. Among the early evidences of the use of these number symbols are notches found on stone columns erected by the Scythian Emperor of India named Asoka, in around 250 BC [1].

The earliest reference to the concept of probability may be traced back to a gambler's manual written by Girolamo Cardano (1501–1576) [2]. However, Blaise Pascal (1623–1662) and Pierre Fermat (1601–1665) were the first two individuals who solved independently and correctly the problem of dividing the winnings in a game of chance [1,2]. Boolean algebra, which plays a pivotal role in modern probability theory, is named after an English mathematician George Boole (1815–1864), who published in 1847 a pamphlet titled "The Mathematical Analysis of Logic: Being an Essay towards a Calculus of Deductive Reasoning" [1–3].

Laplace transforms, frequently used in the area of reliability to find solutions to first-order differential equations, were developed by a French mathematician named Pierre-Simon Laplace (1749–1827). A more detailed history of mathematics and probability is available in [1,2]. This chapter presents basic mathematical concepts that will be useful to understand subsequent chapters of this book.

## 2.2 Arithmetic Mean and Mean Deviation

A given set of data related to reliability/safety of robot systems is useful only if it is analyzed properly. More clearly, there are certain characteristics of the data that are useful to describe the nature of a given data-set, thus making better associated decisions. Thus, this section presents two statistical measures considered useful to study reliability and safety-related data for robot systems [4–6].

### 2.2.1 Arithmetic Mean

This is expressed by

$$m = \frac{\sum_{j=1}^{n} y_j}{n} \tag{2.1}$$

where

$m$ is the mean value (i.e., arithmetic mean),
$y_j$ is the data value $j$, for $j = 1,2,3,\ldots,n$ and
$n$ is the number of data values.

It is to be noted that the arithmetic mean is generally simply called "mean."

**EXAMPLE 2.1**

Assume that the quality control department of a robot system manufacturing company inspected five identical robot systems and discovered 1, 5, 3, 4, and 2 defects in each respective robot system. Calculate the average number of defects (i.e., arithmetic mean) per robot system.

By inserting the given data values into Equation 2.1, we obtain:

$$m = \frac{1+5+3+4+2}{5} = 3 \text{ defects per robot system}$$

Thus, the average number of defects per robot system is three. In other words, the arithmetic mean of the given dataset is 3.

### 2.2.2 Mean Deviation

Mean deviation is a measure of dispersion whose value indicates the degree to which a given set of data tends to spread about a mean value. Mean deviation is expressed by

$$\text{MD} = \frac{\sum_{j=1}^{k} \left| y_j - m \right|}{k} \tag{2.2}$$

where

$k$ is the number of data values,
MD is the mean deviation,
$m$ is the mean value of the given dataset,

$y_j$ is the data value $j$, for $j = 1,2,3,\ldots,k$ and
$|y_j - m|$ is the absolute value of the deviation of $y_j$ from $m$.

**EXAMPLE 2.2**

Calculate the mean deviation of the dataset given in Example 2.1.

In Example 2.1, the calculated mean value (i.e., arithmetic mean) of the given dataset is three defects per robot system. Thus, using this calculated value and the given data values in Equation 2.2, we obtain:

$$\begin{aligned} MD &= \frac{|1-3| + |5-3| + |3-3| + |4-3| + |2-3|}{5} \\ &= \frac{[2+2+0+1+1]}{5} \\ &= 1.2 \end{aligned}$$

Thus, the mean deviation of the dataset in Example 2.1 is 1.2.

## 2.3 Boolean Algebra Laws

Boolean algebra, named after its founder, George Boole (1815–1864), is used to a degree in the studies of reliability and safety-related in robot systems. Some of its laws that are considered useful to understand subsequent chapters of this book are presented below [3–5,7,8].

$$C.D = D.C \tag{2.3}$$

where
C is an arbitrary set or event,
D is an arbitrary set or event and
Dot(.) denotes the intersection of sets.

It is to be noted that sometimes Equation 2.3 is written without the dot (e.g., CD), but it still conveys the same meaning.

$$C + D = D + C \tag{2.4}$$

where
+ is the union of sets or events.

$$C + C = C \tag{2.5}$$

$$CC = C \tag{2.6}$$

$$D(D + C) = D \tag{2.7}$$

$$C + CD = C \tag{2.8}$$

$$C(D + E) = CD + CE \tag{2.9}$$

where
E is an arbitrary set or event.

$$(C + D)(C + E) = C + D\,E \tag{2.10}$$

$$(C + D) + E = C + (D + E) \tag{2.11}$$

$$(CD)E = C(DE) \tag{2.12}$$

It is to be noted that in the published literature, Equations 2.3 and 2.4 are referred to as commutative law, Equations 2.5 and 2.6 as idempotent law, Equations 2.7 and 2.8 as absorption law, Equations 2.9 and 2.10 as distributive law, and Equations 2.11 and 2.12 as associative law [9].

## 2.4 Probability Definition and Properties

The probability is defined as follows [9,10]:

$$P(Z) = \lim_{m \to \infty} \left( \frac{M}{m} \right) \tag{2.13}$$

where
$P(Z)$ is the occurrence probability of event $Z$ and
$M$ is the number of times event $Z$ occurs in the $m$ repeated experiments.

Some of the basic properties of probability are as follows [8–10]:

- The probability of the occurrence of an event, say event $Z$, is

$$0 \le P(Z) \le 1 \tag{2.14}$$

- The probability of the occurrence and nonoccurrence of an event, say event Z, is always:

$$P(Z) + P(\bar{Z}) = 1 \tag{2.15}$$

where

$P(Z)$ is the probability of the occurrence of event $Z$.
$P(\bar{Z})$ is the probability of the nonoccurrence of event $Z$.

- The probability of an intersection of $m$-independent events is

$$P(Z_1 Z_2 Z_3 \ldots Z_m) = P(Z_1)P(Z_2)P(Z_3)\, P(Z_m) \tag{2.16}$$

where

$P(Z_j)$ is the probability of the occurrence of event $Z_j$, for $j = 1,2,3,\ldots,m$.

- The probability of the union of $m$-independent events is

$$P(Z_1 + Z_2 + \cdots + Z_m) = 1 - \prod_{j=1}^{m} (1 - P(Z_j)) \tag{2.17}$$

- The probability of the union of $m$ mutually exclusive events is

$$P(Z_1 + Z_2 + \cdots + Z_m) = \sum_{j=1}^{m} P(Z_j) \tag{2.18}$$

**EXAMPLE 2.3**

Assume that a robot system is composed of two critical subsystems $Z_1$ and $Z_2$. The failure of either subsystem can, directly or indirectly, lead to a serious accident. The probability of failure of subsystems $Z_1$ and $Z_2$ is 0.01 and 0.02, respectively.

Calculate the probability of the occurrence of a serious accident in a robot system if both of these subsystems fail independently.

By inserting the specified data values into Equation 2.17, we obtain:

$$\begin{aligned} P(Z_1 + Z_2) &= 1 - \prod_{j=1}^{2} (1 - P(Z_j)) \\ &= P(Z_1) + P(Z_2) - P(Z_1)P(Z_2) \\ &= 0.01 + 0.02 - (0.01)(0.02) \\ &= 0.0298 \end{aligned}$$

Thus, the occurrence probability of a serious robot system accident is 0.0298.

## 2.5 Probability Distribution-Related Definitions

This section presents three probability distribution-related definitions considered useful to perform various types of reliability and safety studies concerned with robot systems.

### 2.5.1 Cumulative Distribution Function

For a continuous random variable, the cumulative distribution function is expressed by [9,10]

$$F(t) = \int_{-\infty}^{t} f(x)\,dx \tag{2.19}$$

where
$x$ is a continuous random variable,
$f(x)$ is the probability density function, and
$F(t)$ is the cumulative distribution function.

For $t = \infty$, Equation 2.19 becomes

$$F(\infty) = \int_{-\infty}^{\infty} f(x)\,dx = 1 \tag{2.20}$$

It means that the total area under the probability density curve is equal to unity.

Normally, in reliability and safety studies of robot systems, Equation 2.19 is simply written as

$$F(t) = \int_{0}^{t} f(x)\,dx \tag{2.21}$$

**EXAMPLE 2.4**

Assume that the probability (i.e., failure) density function of a robot system is

$$f(t) = \theta e^{-\theta t}, \quad \text{for } t \geq 0, \theta > 0 \tag{2.22}$$

where
$t$ is a continuous random variable (i.e., time),
$\theta$ is robot system failure rate and

$f(t)$ is the probability density function (generally, in the area of reliability-engineering, it is called the failure density function).

Obtain an expression for the robot system cumulative distribution function with the aid of Equation 2.22.

By inserting Equation 2.22 into Equation 2.21, we get

$$F(t) = \int_0^t \theta e^{-\theta t} dt$$
$$= 1 - e^{-\theta t} \tag{2.23}$$

Thus, Equation 2.23 is the expression for the robot system cumulative distribution function.

### 2.5.2 Probability Density Function

For a continuous random variable, the probability density function is defined by [10]

$$f(t) = \frac{dF(t)}{dt} \tag{2.24}$$

where

$f(t)$ is the probability density function and
$F(t)$ is the cumulative distribution function.

**EXAMPLE 2.5**

Prove by using Equation 2.23 that Equation 2.22 is the probability density function.

By substituting Equation 2.23 into Equation 2.24, we get

$$f(t) = \frac{d(1 - e^{-\theta t})}{dt}$$
$$= \theta e^{-\theta t} \tag{2.25}$$

Equations 2.25 and 2.22 are identical.

### 2.5.3 Expected Value

The expected value of a continuous random variable is defined by [10]

$$E(t) = \int_{-\infty}^{\infty} t f(t)\, dt \tag{2.26}$$

where

$E(t)$ is the expected value (i.e., mean value) of the continuous random variable $t$.

**EXAMPLE 2.6**

Find the mean value (i.e., expected value) of the probability (failure) density function expressed by Equation 2.22.

By substituting Equation 2.22 into Equation 2.26, we get

$$
\begin{aligned}
E(t) &= \int_0^{\infty} t\theta e^{-\theta t}\,dt \\
&= \left[-te^{-\theta t}\right]_0^{\infty} - \left[-\frac{e^{-\theta t}}{\theta}\right]_0^{\infty} \\
&= \frac{1}{\theta}
\end{aligned}
\tag{2.27}
$$

Thus, the mean value (i.e., expected value) of the probability (failure) density function expressed by Equation 2.22 is given by Equation 2.27.

## 2.6 Probability Distributions

Although there are a large number of probability or statistical distributions in published literature, this section presents just four such distributions considered useful to perform various types of robot system reliability and safety studies [11–13].

### 2.6.1 Exponential Distribution

This is one of the simplest continuous random variable distributions often used in the industrial sector, particularly to perform reliability-related studies. Its probability density function is defined by [9,14]

$$f(t) = \theta e^{-\theta t}, \quad \text{for } \theta > 0, t \geq 0 \tag{2.28}$$

where

$t$ is the time (i.e., a continuous random variable),
$f(t)$ is the probability density function, and
$\theta$ is the distribution parameter.

By substituting Equation 2.28 into Equation 2.21, we obtain the following expression for the cumulative distribution function:

$$F(t) = 1 - e^{-\theta t} \tag{2.29}$$

Using Equations 2.26 and 2.28, we get the following expression for the distribution expected value (i.e., mean value):

$$E(t) = \frac{1}{\theta} \tag{2.30}$$

**EXAMPLE 2.7**

Assume that the mean time to failure of a robot system is 2000 h. Calculate the robot system's probability of failure during a 1000-h mission by using Equations 2.29 and 2.30.

By substituting the given data value into Equation 2.30, we get

$$\theta = \frac{1}{2000} = 0.0005 \text{ failures per hour}$$

By inserting the calculated and the given data values into Equation 2.29, we obtain:

$$\begin{aligned} F(100) &= 1 - e^{-(0.0005)(1000)} \\ &= 0.3935 \end{aligned}$$

Thus, the robot system's probability of failure during the 1000-h mission is 0.3935.

### 2.6.2 Rayleigh Distribution

This continuous random variable probability distribution is named after its founder, John Raleigh (1842–1919) [1]. The probability density function of the distribution is defined by

$$f(t) = \left(\frac{1}{\gamma^2}\right) t e^{-(t/\gamma)^2}, \quad \text{for } \gamma > 0, t \geq 0 \tag{2.31}$$

where

$\gamma$ is the distribution parameter.

By inserting Equation 2.31 into Equation 2.21, we get the following equation for the cumulative distribution function:

$$F(t) = 1 - e^{-(t/\gamma)^2} \tag{2.32}$$

By substituting Equation 2.31 into Equation 2.26, we obtain the following equation for the distribution expected value:

$$E(t) = \gamma \Gamma\left(\frac{3}{2}\right) \tag{2.33}$$

where

$\Gamma(.)$ is the gamma function and is expressed by

$$\Gamma(m) = \int_0^\infty t^{m-1} e^{-t} dt, \quad \text{for } m > 0 \tag{2.34}$$

### 2.6.3 Weibull Distribution

This continuous random variable distribution was developed by Walliodi Weibull, a Swedish professor in mechanical engineering in the early 1950s [15]. The distribution probability density function is defined by

$$f(t) = \frac{ct^{c-1}}{\gamma^c} e^{-(t/\gamma)^c}, \quad \text{for } \gamma > 0, c > 0, t \geq 0 \tag{2.35}$$

where

$c$ and $\gamma$ are the distribution shape and scale parameters, respectively.

By substituting Equation 2.35 into Equation 2.21, we get the following equation for the cumulative distribution function:

$$F(t) = 1 - e^{-(t/\gamma)^c} \tag{2.36}$$

By inserting Equation 2.35 into Equation 2.26, we get the following equation for the distribution expected value:

$$E(t) = \gamma\Gamma\left(1 + \frac{1}{c}\right) \tag{2.37}$$

It is to be noted that for $c = 2$ and $c = 1$, the Rayleigh and exponential distributions are the special cases of this distribution, respectively.

### 2.6.4 Bathtub Hazard Rate Curve Distribution

This is another continuous random variable distribution and it was developed in 1981 [16]. In the published literature by other authors around the world, it is generally referred to as the Dhillon distribution/law/model [17–36]. The distribution can represent bathtub-shape, increasing, and decreasing hazard rates.

The probability density function of the distribution is defined by [16]

$$f(t) = c\gamma(\gamma t)^{c-1} e^{-\left\{e^{(\gamma t)^c} - (\gamma t)^c - 1\right\}}, \quad \text{for } t \geq 0, \gamma > 0, c > 0, \tag{2.38}$$

where
$c$ and $\gamma$ are the distribution shape and scale parameters, respectively.

By inserting Equation 2.38 into Equation 2.27, we get the following equation for the cumulative distribution function:

$$F(t) = 1 - e^{-\left\{e^{(\gamma t)^c} - 1\right\}} \tag{2.39}$$

It is to be noted that for $c = 0.5$, this probability distribution gives the bathtub-shaped hazard rate curve, and for $c = 1$ it gives the extreme value probability distribution. More clearly, the extreme value probability distribution is the special case of this probability distribution at $c = 1$.

## 2.7 Laplace Transform Definition, Common Laplace Transforms, and Final-Value Theorem Laplace Transform

This section presents various aspects of Laplace transforms considered useful to conduct robot system reliability and safety-related studies.

### 2.7.1 Laplace Transform Definition

The Laplace transform (named after a French mathematician, Pierre-Simon Laplace [1749–1827]) of a function, say $f(t)$, is defined by [1,37,38]

$$f(s) = \int_0^\infty f(t)e^{-st}dt \tag{2.40}$$

where
$t$ is a variable,
$s$ is the Laplace transform variable, and
$f(s)$ is the Laplace transform of function $f(t)$.

**EXAMPLE 2.8**

Obtain the Laplace transform of the following function:

$$f(t) = e^{-\lambda t} \tag{2.41}$$

where
 $\lambda$ is a constant.

By substituting Equation 2.41 into Equation 2.40, we obtain:

$$\begin{aligned} f(s) &= \int_0^\infty e^{-\lambda t} e^{-st}\,dt \\ &= \frac{e^{-(s+\lambda)t}}{(s+\lambda)}\Big|_0^\infty \\ &= \frac{1}{s+\lambda} \end{aligned} \tag{2.42}$$

Thus, Equation 2.42 is the Laplace transform of Equation 2.41.

### 2.7.2 Laplace Transforms of Common Functions

Laplace transforms of some commonly occurring functions in robot system reliability and safety-related analysis studies are presented in Table 2.1 [37–39].

**TABLE 2.1**

Laplace Transforms of Some Functions

| No. | $f(t)$ | $f(s)$ |
|---|---|---|
| 1 | $t^n$, for $n = 0,1,2,3\ldots$ | $\frac{n!}{s^{n+1}}$ |
| 2 | $K$ (a constant) | $\frac{K}{s}$ |
| 3 | $t$ | $\frac{1}{s^2}$ |
| 4 | $e^{-\lambda t}$ | $\frac{1}{s+\lambda}$ |
| 5 | $te^{-\lambda t}$ | $\frac{1}{(s+\lambda)^2}$ |
| 6 | $t\,f(t)$ | $-\frac{df(s)}{ds}$ |
| 7 | $\frac{df(t)}{dt}$ | $s\,f(s) - f(0)$ |
| 8 | $\theta_1 f_1(t) + \theta_2 f_2(t)$ | $\theta_1 f_1(s) + \theta_2 f_2(s)$ |

### 2.7.3 Final Value Theorem Laplace Transform

If the following limits exist, then the final value theorem may be stated as

$$\lim_{t\to\infty} f(t) = \lim_{s\to 0}[sf(s)] \tag{2.43}$$

**EXAMPLE 2.9**

Prove, by using the following equation, that the left-hand side of Equation 2.43 is equal to its right side:

$$f(t) = \frac{\theta_1}{(\theta_1+\theta_2)} + \frac{\theta_2}{(\theta_1+\theta_2)} e^{-(\theta_1+\theta_2)t} \tag{2.44}$$

where
$\theta_1$ and $\theta_2$ are the constants.

By substituting Equation 2.44 into the left-hand side of Equation 2.43, we get

$$\lim_{t\to\infty}\left[\frac{\theta_1}{(\theta_1+\theta_2)} + \frac{\theta_2}{(\theta_1+\theta_2)} e^{-(\theta_1+\theta_2)t}\right] = \frac{\theta_1}{\theta_1+\theta_2} \tag{2.45}$$

With the aid of Table 2.1, we obtain the following Laplace transforms of Equation 2.44:

$$f(s) = \frac{\theta_1}{s(\theta_1+\theta_2)} + \frac{\theta_2}{(\theta_1+\theta_2)} \frac{1}{(s+\theta_1+\theta_2)} \tag{2.46}$$

By inserting Equation 2.46 into the right-hand side of Equation 2.43, we obtain:

$$\lim_{s\to 0} s\left[\frac{\theta_1}{s(\theta_1+\theta_2)} + \frac{\theta_2}{(\theta_1+\theta_2)} \frac{1}{(s+\theta_1+\theta_2)}\right] = \frac{\theta_1}{\theta_1+\theta_2} \tag{2.47}$$

As the right-hand sides of Equations 2.45 and 2.47 are identical, it proves that the left-hand side of Equation 2.43 is equal to its right side.

## 2.8 Solving First-Order Differential Equations Using Laplace Transforms

Normally, Laplace transforms are used to find solutions to first-order linear differential equations in reliability and safety analysis-related studies of robot systems. The example presented below demonstrates the finding of solutions to a set of linear first-order differential equations, describing a robot system with respect to reliability, using Laplace transforms.

**EXAMPLE 2.10**

Assume that a robot system can be in any of the three states: operating normally, failed due to a hardware failure, or failed due to a human error. The following three first-order linear differential equations describe the robot system under consideration:

$$\frac{dP_0(t)}{dt} + (\lambda_1 + \lambda_2)P_0(t) = 0 \tag{2.48}$$

$$\frac{dP_1(t)}{dt} - \lambda_1 P_0(t) = 0 \tag{2.49}$$

$$\frac{dP_2(t)}{dt} - \lambda_2 P_0(t) = 0 \tag{2.50}$$

where

$\lambda_1$ is the robot system constant hardware failure rate,

$\lambda_2$ is the robot system constant human error rate, and

$P_j(t)$ is the probability that the robot system is in state $j$ at time $t$, for $j = 0$ (operating normally), $j = 1$ (failed due to a hardware failure), and $j = 2$ (failed due to a human error),

At time $t = 0$, $P_0(0) = 1$, $P_1(0) = 0$, and $P_2(0) = 0$.

Solve differential Equations 2.48 through 2.50 with the aid of Laplace transforms.

With the aid of Table 2.1, differential Equations 2.48 through 2.50, and the given initial conditions, we get

$$sP_0(s) - 1 + (\lambda_1 + \lambda_2)P_0(s) = 0 \tag{2.51}$$

$$sP_1(s) - \lambda_1 P_0(s) = 0 \tag{2.52}$$

$$sP_2(s) - \lambda_2 P_0(s) = 0 \tag{2.53}$$

By solving Equations 2.51 through 2.53, we obtain:

$$P_0(s) = \frac{1}{(s + \lambda_1 + \lambda_2)} \tag{2.54}$$

$$P_1(s) = \frac{\lambda_1}{s(s + \lambda_1 + \lambda_2)} \tag{2.55}$$

$$P_2(s) = \frac{\lambda_2}{s(s + \lambda_1 + \lambda_2)} \tag{2.56}$$

By taking the inverse Laplace transforms of Equations 2.54 through 2.56, we get

$$P_0(t) = e^{-(\lambda_1+\lambda_2)t} \tag{2.57}$$

$$P_1(t) = \frac{\lambda_1}{(\lambda_1 + \lambda_2)}\left[1 - e^{-(\lambda_1+\lambda_2)t}\right] \tag{2.58}$$

$$P_2(t) = \frac{\lambda_2}{(\lambda_1 + \lambda_2)}\left[1 - e^{-(\lambda_1+\lambda_2)t}\right] \tag{2.59}$$

Thus, Equations 2.57 through 2.59 are the solutions to differential Equations 2.48 through 2.50.

## 2.9 Problems

1. A robot system manufacturing company's quality control department inspected seven identical robot systems and found 4, 7, 8, 3, 2, 1, and 5 defects in each respective robot system. Calculate the average number of defects (i.e., the arithmetic mean) per robot system.
2. Calculate the mean deviation of the dataset given in Question 1.
3. What is commutative law?
4. Define probability.
5. What are the basic properties of probability?
6. Define expected value of a continuous random variable.
7. Define the following items:
   a. Probability density function; and
   b. Cumulative distribution function.
8. Write down the probability density function for the Rayleigh distribution.
9. Prove Equations 2.57 through 2.59 by using Equations 2.54 through 2.56.
10. What are the special case distributions of the Wiebull and bathtub hazard rate curve distributions?

## References


1. Eves, H., *An Introduction to the History of Mathematics*, Holt, Rinehart and Winston, New York, 1976.
2. Owen, D.B., Editor, *On the History of Statistics and Probability*, Marcel Dekker, New York, 1976.
3. Lipschutz, S., *Set Theory*, McGraw Hill Book Company, New York, 1964.
4. Speigel, M.R., *Statistics*, McGraw Hill Book Company, New York, 1961.
5. Speigel, M.R., *Probability and Statistics*, McGraw Hill Book Company, New York, 1975.
6. Dhillon, B.S., *Reliability, Quality, and Safety for Engineers*, CRC Press, Boca Raton, Florida, 2004.
7. Fault Tree Handbook, Report No. NUREG-0492, U.S. Nuclear Regulatory Commission, Washington, DC, 1981.
8. Lipschutz, S., *Probability*, McGraw Hill Book Company, New York, 1965.
9. Dhillon, B.S., *Computer System Reliability: Safety and Usability*, CRC Press, Boca Raton, Florida, 2013.
10. Mann, N.R., Schafer, R.E., Singpurwalla, N.P., *Methods for Statistical Analysis of Reliability and Life Data*, John Wiley and Sons, New York, 1974.
11. Shooman, M.L., *Probabilistic Reliability: An Engineering Approach*, McGraw Hill Book Company, New York, 1968.
12. Patel, J.K., Kapadia, C.H., Owen, D.H., *Handbook of Statistical Distributions*, Marcel Dekker, New York, 1976.
13. Dhillon, B.S., *Reliability Engineering in Systems Design and Operation*, Van Nostrand Reinhold, New York, 1983.
14. Davis, D.J., Analysis of some failure data, *J. Am. Stat. Assoc.*, 1952, pp. 113–150.
15. Weibull, W., A statistical distribution function of wide applicability, *J. Appl. Mech.*, Vol. 18, 1951, pp. 293–297.
16. Dhillon, B.S., Life distributions, *IEEE Transactions on Reliability*, Vol. 30, 1981, pp. 457–460.
17. Baker, R.D., Non-parametric estimation of the renewal function, *Computers Operations Research*, Vol. 20, No. 2, 1993, pp. 167–178.
18. Cabana, A., Cabana, E.M., Goodness-of-fit to the exponential distribution, focused on Weibull alternatives, *Communications in Statistics-Simulation and Computation*, Vol. 34, 2005, pp. 711–723.
19. Grane, A., Fortiana, J., A directional test of exponentiality based on maximum correlations, *Metrika*, Vol. 73, 2011, pp. 255–274.
20. Henze, N., Meintnis, S.G., Recent and classical tests for exponentiality: A partial review with comparisons, *Metrika*, Vol. 61, 2005, pp. 29–45.
21. Jammalamadaka, S.R., Taufer, E., Testing exponentiality by comparing the empirical distribution function of the normalized spacings with that of the original data, *Journal of Nonparametric Statistics*, Vol. 15, No. 6, 2003, pp. 719–729.
22. Hollander, M., Laird, G., Song, K.S., Non-parametric interference for the proportionality function in the random censorship model, *Journal of Nonparametric Statistics*, Vol. 15, No. 2, 2003, pp. 151–169.

23. Jammalamadaka, S.R., Taufer, E., Use of mean residual life in testing departures from exponentiality, *Journal of Nonparametric Statistics*, Vol. 18, No. 3, 2006, pp. 277–292.
24. Kunitz, H., Pamme, H., The mixed gamma ageing model in life data analysis, *Statistical Papers*, Vol. 34, 1993, pp. 303–318.
25. Kunitz, H., A new class of bathtub-shaped hazard rates and its application in comparison of two test-statistics, *IEEE Transactions on Reliability*, Vol. 38, No. 3, 1989, pp. 351–354.
26. Meintanis, S.G., A class of tests for exponentiality based on a continuum of moment conditions, *Kybernetika*, Vol. 45, No. 6, 2009, pp. 946–959.
27. Morris, K., Szynal, D., Goodness-of-fit tests based on characterizations involving moments of order statistics, *International Journal of Pure and Applied Mathematics*, Vol. 38, No. 1, 2007, pp. 83–121.
28. Na, M.H., Spline hazard rate estimation using censored data, *Journal of KSIAM*, Vol. 3, No. 2, 1999, pp. 99–106.
29. Morris, K., Szynal, D., Some U-statistics in goodness-of-fit tests derived from characterizations via record values, *International Journal of Pure and Applied Mathematics*, Vol. 46, No. 4, 2008, pp. 339–414.
30. Nam, K.H., Park, D.H., Failure rate for Dhillon model, Proceedings of the Spring Conference of the Korean Statistical Society, 1997, pp. 114–118.
31. Nimoto, N., Zitikis, R., The Atkinson index, the Moran statistic, and testing exponentiality, *Journal of the Japan Statistical Society*, Vol. 38, No. 2, 2008, pp. 187–205.
32. Nam, K.H., Chang, S.J., Approximation of the renewal function for Hjorth model and Dhillon model, *Journal of the Korean Society for Quality Management*, Vol. 34, No. 1, 2006, pp. 34–39.
33. Noughabi, H.A., Arghami, N.R., Testing exponentiality based on characterizations of the exponential distribution, *Journal of Statistical Computation and Simulation*, Vol. 1, First, 2011, pp. 1–11.
34. Szynal, D., Goodness-of-fit tests derived from characterizations of continuous distributions, *Stability in Probability*, Banach Center Publications, Vol. 90, Institute of Mathematics, Polish Academy of Sciences, Warszawa, Poland, 2010, pp. 203–223.
35. Szynal, D., Wolynski, W., Goodness-of-fit tests for exponentiality and Rayleigh distribution, *International Journal of Pure and Applied Mathematics*, Vol. 78, No. 5, 2012, pp. 751–772.
36. Nam, K.H., Park, D.H., A study on trend changes for certain parametric families, *Journal of the Korean Society for Quality Management*, Vol. 23, No. 3, 1995, pp. 93–101.
37. Oberhettinger, F., Badii, L., *Tables of Laplace Transforms*, Springer-Verlag, Inc., New York, 1973.
38. Spiegel, M.R., *Laplace Transforms*, McGraw Hill Book Company, New York, 1965.
39. Nixon, F.E., *Handbook of Laplace Transformation: Fundamentals, Applications, Tables, and Examples*, Prentice-Hall, Inc., Englewood Cliffs, New Jersey, 1960.

# 3

# Reliability and Safety Basics

## 3.1 Introduction

The history of the field of reliability goes back to the early years of the 1930s, when probability concepts were applied to problems in the area of electric power generation. However, the real beginning of the field is normally regarded as World War II, when German scientists applied basic reliability concepts for improving the reliability of their V1 and V2 rockets [1–4]. Today, the field of reliability has become a well-developed discipline and has branched into specialized areas such as power system reliability, human reliability, software reliability, and mechanical reliability [3,4].

The history of the field of safety may be traced back to 1868, when a patent for a barrier safeguard was awarded in the United States [5]. Twenty-five years later, in 1893, the US Congress passed the Railway Safety Act [5]. Today, the safety field has branched out into many specialized areas, including patient safety, workplace safety, and system safety [6].

This chapter presents various important aspects of reliability and safety basics considered useful to understand subsequent chapters of this book, as well as for the robotics sector at large.

## 3.2 Bathtub Hazard Rate Curve

Usually, this curve is used to describe the failure rate of engineering systems/items and is shown in Figure 3.1. The curve is called the bathtub hazard rate curve because it resembles the shape of a bathtub.

As shown in Figure 3.1, the curve is divided into three sections—the burn-in period, useful-life period, and wear-out period. During the burn-in period, the system/item hazard rate decreases with time *t*; some of the reasons for the occurrence of failures during this period are poor manufacturing methods and processes, substandard materials and workmanship, poor quality control, human error, and inadequate debugging [4,7]. Other terms

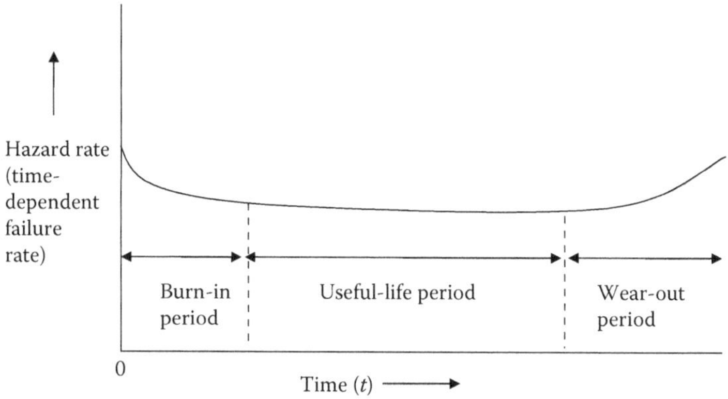


**FIGURE 3.1**
Bathtub hazard rate curve.

used in the published literature for this decreasing-hazard-rate region are "infant-mortality region," "break-in region," and "debugging region."

During the useful-life period, the hazard rate remains constant. Some of the causes for the occurrence of failures in this region are as follows [4,7]:

- Undetectable defects
- Low safety factors
- Human errors
- Higher random stress than expected
- Abuse
- Natural failures

Finally, during the wear-out period, the hazard rate increases with time $t$ due to reasons such as wear from aging, poor maintenance practices, wrong overhaul practices, short designed-in life of the system/item under consideration, and wear due to friction, corrosion, and creep [4,7].

Mathematically, the following equation can be used to represent the bathtub hazard rate curve shown in Figure 3.1 [8]:

$$\lambda(t) = \theta a(\theta t)^{a-1} e^{(\theta t)^a} \tag{3.1}$$

where
$t$ is time,
$\lambda(t)$ is hazard rate (time-dependent failure rate),
$\theta$ is the scale parameter, and
$a$ is the shape parameter.

Equation 3.1 gives the shape of the bathtub hazard rate curve shown in Figure 3.1, at $a = 0.5$.

## 3.3 General Reliability-Related Formulas

A number of general formulas are often used to perform various types of reliability analysis. Four of these formulas that are based on the reliability function are presented below.

### 3.3.1 Probability (or Failure) Density Function

The probability (or failure) density function is expressed by [4]

$$f(t) = -\frac{dR(t)}{dt} \tag{3.2}$$

where
$R(t)$ is the system/item reliability at time $t$ and
$f(t)$ is the probability (or failure) density function.

**EXAMPLE 3.1**

Assume that a robot system's reliability is expressed by

$$R_{rs}(t) = e^{-\lambda_{rs} t} \tag{3.3}$$

where
$R_{rs}(t)$ is the robot system reliability at time $t$ and
$\lambda_{rs}$ is the robot system constant failure rate.

Obtain an expression for the probability (or failure) density function of the robot system with the aid of Equation 3.2.

By substituting Equation 3.3 into Equation 3.2, we get

$$f(t) = -\frac{de^{-\lambda_{rs} t}}{\lambda_{rs} e^{-\lambda_{rs} t}} \tag{3.4}$$

Thus, Equation 3.4 is the expression for the probability (or failure) density function of the robot system.

### 3.3.2 Hazard Rate (or Time-Dependent Failure Rate) Function

This is expressed by

$$\lambda(t) = \frac{f(t)}{R(t)} \tag{3.5}$$

where

$\lambda(t)$ is the system/item hazard rate (or time-dependent failure rate).

By substituting Equation 3.2 into Equation 3.5, we obtain:

$$\lambda(t) = -\frac{1}{R(t)} \cdot \frac{dR(t)}{dt} \tag{3.6}$$

**EXAMPLE 3.2**

Obtain an expression for the hazard rate of the robot system with the aid of Equations 3.3 and 3.6 and comment on the final result. By inserting Equation 3.3 into Equation 3.6, we obtain:

$$\begin{aligned} \lambda_{rs}(t) &= -\frac{1}{e^{-\lambda_{rs}t}} \cdot \frac{de^{-\lambda_{rs}t}}{dt} \\ &= \lambda_{rs} \end{aligned} \tag{3.7}$$

Thus, the hazard rate of the robot system is given by Equation 3.7. The right-hand side of this equation is not a function of time $t$. In other words, it is constant. Generally, it is referred to as the constant failure rate of a system/item (in this case, of the robot system) because it does not depend on time $t$.

### 3.3.3 General Reliability Function

This can be obtained by using Equation 3.6. Thus, using Equation 3.6, we get

$$-\lambda(t)dt = -\frac{1}{R(t)} \cdot dR(t) \tag{3.8}$$

By integrating both sides of Equation 3.8 over the time interval [0,$t$], we get

$$-\int_0^t \lambda(t)dt = \int_1^{R(t)} \frac{1}{R(t)} \cdot dR(t) \tag{3.9}$$

Since, at $t = 0$, $R(t) = 1$.

Evaluating the right-hand side of Equation 3.9 and then rearranging it yields

$$\ln R(t) = -\int_0^t \lambda(t)dt \tag{3.10}$$

Thus, from Equation 3.10, we get

$$R(t) = e^{-\int_0^t \lambda(t)dt} \tag{3.11}$$

Thus, Equation 3.11 is the general reliability function. This equation can be used to obtain the reliability function of a system/item when its times to failure follow any time-continuous probability distribution (e.g., exponential, Rayleigh, and Weibull).

**EXAMPLE 3.3**

Assume that the hazard rate of a robot system is expressed by Equation 3.1. Obtain an expression for the reliability function of the robot system with the aid of Equation 3.11.

By inserting Equation 3.1 into Equation 3.11, we get

$$\begin{aligned} R(t) &= e^{-\int_0^t \{\theta a(\theta t)^{a-1} e^{(\theta t)^a}\}dt} \\ &= e^{-\{e^{(\theta t)^a}-1\}} \end{aligned} \tag{3.12}$$

Thus, Equation 3.12 is the expression for the reliability function of the robot system.

### 3.3.4 Mean Time to Failure

The mean time to failure of a system/item can be obtained by using any of the following three formulas [9,10]:

$$MTTF = \int_0^\infty R(t)dt \tag{3.13}$$

or

$$MTTF = \lim_{s \to 0} R(s) \tag{3.14}$$

or

$$MTTF = E(t) = \int_0^\infty tf(t)dt \tag{3.15}$$

where
$MTTF$ is the mean time to failure of a system/item,
$E(t)$ is the expected value,
$s$ is the Laplace transform variable, and
$R(s)$ is the Laplace transform of the reliability function $R(t)$.

**EXAMPLE 3.4**

Using Equation 3.3, prove that Equations 3.13 and 3.14 yield the same result for the robot system mean time to failure.

By inserting Equation 3.3 into Equation 3.13, we get

$$\begin{aligned} MTTF_{rs} &= \int_0^\infty e^{-\lambda_{rs} t} dt \\ &= \frac{1}{\lambda_{rs}} \end{aligned} \tag{3.16}$$

where
$MTTF_{rs}$ is the robot system mean time to failure.

By taking the Laplace transform of Equation 3.3, we obtain

$$\begin{aligned} R_{rs}(s) &= \int_0^\infty e^{-st} e^{-\lambda_{rs} t} dt \\ &= \frac{1}{(s + \lambda_{rs})} \end{aligned} \tag{3.17}$$

where
$R_{rs}(s)$ is the Laplace transform of the robot system reliability function $R_{rs}(t)$.

By substituting Equation 3.17 into Equation 3.14, we get

$$\begin{aligned} MTTF_{rs} &= \lim_{s \to 0} \frac{1}{(s + \lambda_{rs})} \\ &= \frac{1}{\lambda_{rs}} \end{aligned} \tag{3.18}$$

As Equations 3.16 and 3.18 are identical, it proves that Equations 3.13 and 3.14 yield the same result for the robot system mean time to failure.

## 3.4 Reliability Configurations

In analyzing reliability, the analyst can encounter parts of a robot system forming a variety of different configurations. Thus, this section is concerned with the reliability-evaluation of such commonly occurring configurations or networks.

### 3.4.1 Series Network

This is the simplest network or configuration of reliability, and its block diagram is shown in Figure 3.2. The diagram represents an $n$-unit system, and each block in the diagram denotes a unit/part. If any one of the n units/parts fails, the series system/network/configuration fails. In other words, for the successful operation of the series system, all the $n$ system units must operate normally.

The series system, reliability shown in Figure 3.2, reliability is expressed by

$$R_s = P(E_1E_2E_3 \ldots E_n) \tag{3.19}$$

where

$E_j$ is the successful operation (i.e., success event) of unit $j$, for $j = 1,2,3,\ldots,n$.
$P(E_1E_2E_3\ldots E_n)$ is the occurrence probability of events $E_1E_2E_3\ldots E_n$.
$R_s$ is the series system reliability.

For independently failing units, Equation 3.19 becomes

$$R_s = P(E_1)P(E_2)P(E_3)\ldots P(E_n) \tag{3.20}$$

where

$P(E_j)$ is the probability of occurrence of event $E_j$; for $j = 1,2,3,\ldots,n$.

If we let $R_j = P(E_j)$ for $j = 1,2,3,\ldots,n$, Equation 3.20 becomes

$$R_s = R_1R_2R_3 \ldots R_n$$
$$= \prod_{j=1}^{n} R_j \tag{3.21}$$

where

$R_j$ is the unit $j$ reliability, for $j = 1,2,3,\ldots,n$.

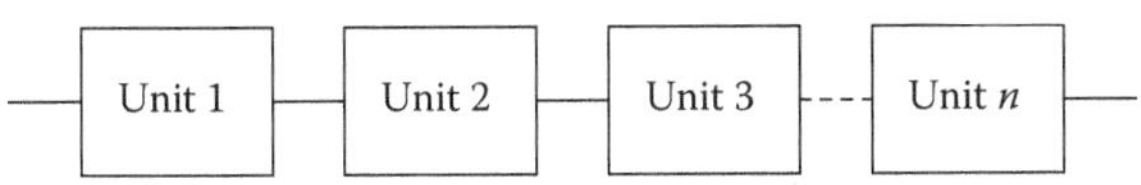


**FIGURE 3.2**
An $n$-unit series system/network/configuration.

For constant failure rate, $\lambda_j$, of unit $j$ from Equation 3.11, we get

$$\begin{aligned} R_j(t) &= e^{-\int_0^t \lambda_j dt} \\ &= e^{-\lambda_j t} \end{aligned} \tag{3.22}$$

where
$R_j(t)$ is the reliability of unit $j$ at time $t$.

By substituting Equation 3.22 into Equation 3.21, we obtain

$$R_s(t) = e^{-\sum_{j=1}^{n} \lambda_j t} \tag{3.23}$$

where
$R_s(t)$ is the series system reliability at time $t$.

Substituting Equation 3.23 into Equation 3.13 yields the following expression for the series-system's mean time to failure:

$$\begin{aligned} MTTF_s &= \int_0^\infty e^{-\sum_{j=1}^{n} \lambda_j t} dt \\ &= \frac{1}{\sum_{j=1}^{n} \lambda_j} \end{aligned} \tag{3.24}$$

where
$MTTF_s$ is the series-system's mean time to failure.

By inserting Equation 3.23 into Equation 3.6, we get the following expression for the series-system's hazard rate:

$$\begin{aligned} \lambda_s(t) &= -\frac{1}{e^{-\sum_{j=1}^{n} \lambda_j t}} \left[ -\sum_{j=1}^{n} \lambda_j \right] e^{-\sum_{j=1}^{n} \lambda_j t} \\ &= \sum_{j=1}^{n} \lambda_j \end{aligned} \tag{3.25}$$

where
$\lambda_s(t)$ is the series-system's hazard rate.

It is to be noted that the right-hand side of Equation 3.25 is independent of time $t$. Thus, the left-hand side of this equation is simply $\lambda_s$, the failure rate of the series system. It means that whenever we add up failure rates of independent items/units/parts, we automatically assume that these items/units/parts form a series network or configuration, a worst-case design scenario in regard to reliability.

**EXAMPLE 3.5**

Assume that a robot system is composed of four independent and identical subsystems and the constant failure rate of each subsystem is 0.0001 failures per hour. All four subsystems must operate normally for the robot system to function successfully.

Calculate the robot system's reliability for a 50-h mission, mean time to failure, and failure rate.

By inserting the given data values into Equation 3.23, we obtain:

$$R_s(50) = e^{-((0.0001+0.0001+0.0001+0.0001)(50))}$$
$$= 0.9802$$

Substituting the given data values into Equation 3.24 yields:

$$MTTF_s = \frac{1}{(0.0001 + 0.0001 + 0.0001 + 0.0001)}$$
$$= 2500 \text{ h}$$

By substituting the given data values into Equation 3.25, we get

$$\lambda_s = (0.0001 + 0.0001 + 0.0001 + 0.0001)$$
$$= 0.0004 \text{ failures per hour}$$

Thus, the robot system's reliability, mean time to failure, and failure rate are 0.9802, 2500 h, and 0.0004 failures per hour, respectively.

### 3.4.2 Parallel Network

This network represents a system with $n$ units/items operating simultaneously. For the successful operation of the system, at least one of the $n$ units must operate normally. The block diagram of an $n$-unit parallel system is shown in Figure 3.3; each block in the diagram represents a unit.

The probability of failure of the parallel system shown in Figure 3.3 is expressed by

$$F_p = P(\bar{y}_1\bar{y}_2\bar{y}_3 \ldots \bar{y}_n) \tag{3.26}$$

where

$F_p$ is the probability of failure of the parallel system.

$\bar{y}_j$ is the failure (i.e., failure event) of unit $j$, for $j = 1,2,3,\ldots,n$.

$P(\bar{y}_1\bar{y}_2\bar{y}_3 \ldots \bar{y}_n)$ is the probability of occurrence of events $\bar{y}_1, \bar{y}_2, \bar{y}_3, \ldots,$ and $\bar{y}_n$.

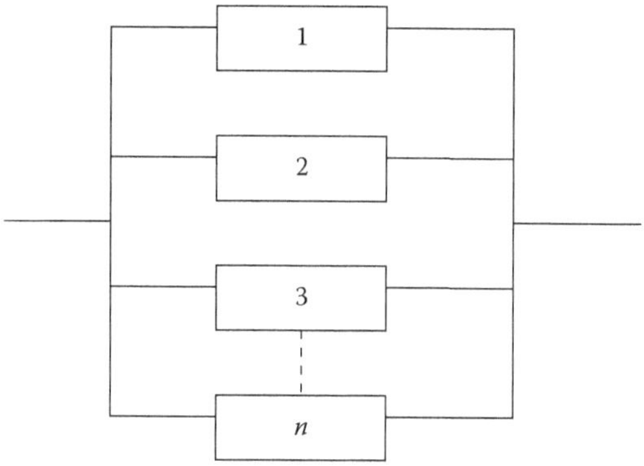


**FIGURE 3.3**
Block diagram of a parallel system or network with $n$ units.

For independently failing parallel units, Equation 3.26 is written as

$$F_p = P(\bar{y}_1)P(\bar{y}_2)P(\bar{y}_3)\ldots P(\bar{y}_n) \tag{3.27}$$

where
$P(\bar{y}_j)$ is the probability of occurrence of failure event $\bar{y}_j$ for $j = 1,2,3,\ldots,n$.

If we let $F_j = P(\bar{y}_j)$, for $j = 1,2,3,\ldots,n$, then Equation 3.27 becomes

$$F_p = \prod_{j=1}^{n} F_j \tag{3.28}$$

where
$F_j$ is the unit $j$ failure probability, for $j = 1,2,3,\ldots,n$.

By subtracting Equation 3.28 from unity, we get

$$R_p = 1 - \prod_{j=1}^{n} F_j \tag{3.29}$$

where
$R_p$ is the parallel system/network reliability.

For constant failure rate $\lambda_j$ of unit $j$, subtracting Equation 3.22 from unity and then inserting it into Equation 3.29 yields:

$$R_p(t) = 1 - \prod_{j=1}^{n} (1 - e^{\lambda_j t}) \tag{3.30}$$

where
$R_p(t)$ is the parallel system/network reliability at time $t$.

For identical units, Equation 3.30 becomes

$$R_p(t) = 1 - (1 - e^{\lambda t})^n \tag{3.31}$$

where
$\lambda$ is the unit constant failure rate.

By substituting Equation 3.31 into Equation 3.13, we get the following expression for the parallel system or network mean time to failure:

$$\begin{aligned} MTTF_p &= \int_0^\infty \left[1 - (1 - e^{-\lambda t})^n\right] dt \\ &= \frac{1}{\lambda} \sum_{j=1}^{n} \frac{1}{j} \end{aligned} \tag{3.32}$$

where
$MTTF_p$ is the identical unit parallel system/network mean time to failure.

**EXAMPLE 3.6**

Assume that a robot system is composed of three identical, active, and independent units. At least one of the units must function normally for the robot system to operate successfully. The failure rate of a unit is 0.0002 failures per hour.

Calculate the robot system's reliability for a 300-h mission and mean time to failure.

By inserting the given data values into Equation 3.31, we obtain:

$$\begin{aligned} R_p(300) &= 1 - [1 - e^{(0.0002)(300)}]^3 \\ &= 0.9998 \end{aligned}$$

Substituting the specified data values into Equation 3.32 yields:

$$\begin{aligned} MTTF_p &= \frac{1}{(0.0002)}\left[1 + \frac{1}{2} + \frac{1}{3}\right] \\ &= 9166.67 \text{ h} \end{aligned}$$

Thus, the robot system's reliability and mean time to failure are 0.9998 and 9166.67 h, respectively.

### 3.4.3 *k*-out-of-*n* Network

In this case, the network/system is composed of $n$ active units, and at least $k$ units out of $n$ active units must operate normally for the successful system operation. The block diagram of a $k$-out-of-$n$ unit network/system is shown in Figure 3.4 and each block in the diagram denotes a unit.

The parallel and series networks are special cases of this network for $k = 1$ and $k = n$, respectively.

With the aid of binomial distribution, for identical and independent units, we have the following expression for reliability of $k$-out-of-$n$ unit network shown in Figure 3.4:

$$R_{k/n} = \sum_{j=k}^{n} \binom{n}{j} R^j (1 - R)^{n-j} \tag{3.33}$$

where

$$\binom{n}{j} = \frac{n!}{(n-j)!j!} \tag{3.34}$$

$R_{k/n}$ is the $k$-out-of-$n$ network/system reliability, and
$R$ is the unit reliability.

For constant failure rates of the identical units, with the aid of Equations 3.11 and 3.33, we get

$$R_{k/n}(t) = \sum_{j=k}^{n} \binom{n}{j} e^{-j\lambda t} (1 - e^{\lambda t})^{n-j} \tag{3.35}$$

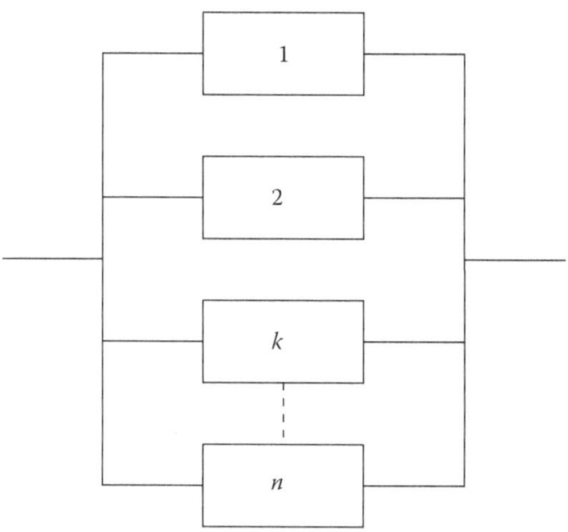


**FIGURE 3.4**
Block diagram of a $k$-out-of-$n$ unit network.

where
$R_{k/n}(t)$ is the $k$-out-of-$n$ network/system reliability at time $t$, and
$\lambda$ is the unit constant failure rate.

By inserting Equation 3.35 into Equation 3.13, we obtain:

$$MTTF_{k/n} = \int_0^\infty \left[ \sum_{j=k}^{n} \binom{n}{j} e^{-j\lambda t} (1 - e^{-\lambda t})^{n-j} \right] dt$$
$$= \frac{1}{\lambda} \sum_{j=k}^{n} \frac{1}{j} \tag{3.36}$$

where
$MTTF_{k/n}$ is the $k$-out-of-$n$ network/system mean time to failure.

**EXAMPLE 3.7**

A robot system has four active, identical, and independent units in parallel. At least two units must operate normally for the successful operation of the robot system. Calculate the robot system's mean time to failure if the unit constant failure rate is 0.0008 failures per hour.

By inserting the given data values into Equation 3.36, we get

$$MTTF_{2/4} = \frac{1}{(0.0008)} \left[ \frac{1}{2} + \frac{1}{3} + \frac{1}{4} \right]$$
$$= 1354.17 \text{ h}$$

Thus, the robot system mean time to failure is 1354.17 h.

### 3.4.4 Standby System

This is another reliability network/system/configuration in which only one unit works and m units are kept in their standby mode. The system contains a total of $(m + 1)$ units, and as soon as the operating unit fails, the switching mechanism detects the failure and turns on one of the standby units. The system fails when all its standby units fail.

Figure 3.5 shows the block diagram of a standby system with one operating and m standby units. Each block in the diagram denotes a unit. With the aid of Figure 3.5, for independent and identical units, perfect switching mechanism and standby units, and time-dependent unit failure rate, we have the following expression for reliability of the standby system [4,11]:

$$R_{ss}(t) = \frac{\sum_{j=0}^{m} \left[ \left[ \int_0^t \lambda(t)dt \right]^j e^{-\int_0^t \lambda(t)dt} \right]}{j!} \tag{3.37}$$

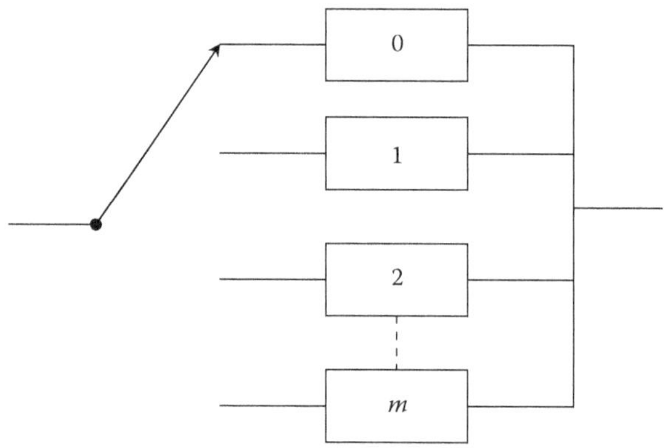

**FIGURE 3.5**
Block diagram of a standby system with one operating and $m$ standby units.

where
$m$ is the number of standby units,
$\lambda(t)$ is the unit time-dependent failure rate or hazard rate, and
$R_{ss}(t)$ is the standby system reliability at time $t$.

For constant unit failure rate (i.e., $\lambda(t) = \lambda$), Equation 3.37 becomes

$$R_{ss}(t) = \frac{\sum_{j=0}^{m} (\lambda t)^j e^{-\lambda t}}{j!} \tag{3.38}$$

where
$\lambda$ is the unit constant failure rate.

By substituting Equation 3.38 into Equation 3.13, we get

$$MTTF_{ss} = \int_0^{\infty} \left[ \frac{\sum_{j=0}^{m} (\lambda t)^j e^{-\lambda t}}{j!} \right] dt$$
$$= \frac{(m+1)}{\lambda} \tag{3.39}$$

where
$MTTF_{ss}$ is the standby system's mean time to failure.

**EXAMPLE 3.8**

A robot system is composed of a standby system having two independent and identical units: one operating, the other on standby. The unit constant failure rate is 0.0004 failures per hour.

Calculate the standby system's reliability for a 100-h mission if the switching mechanism is perfect and the standby unit remains as good as new in its standby mode.

By substituting the specified data values into Equation 3.38, we obtain:

$$R_{ss}(100) = \sum_{j=0}^{1} \frac{\left[\{(0.0004)(100)\}^{j} e^{-(0.0004)(100)}\right]}{j!}$$
$$= 0.9992$$

Thus, the standby system reliability is 0.9992.

### 3.4.5 Bridge Network

Sometimes units or parts in robot systems may form a bridge network, as shown in Figure 3.6. Each block in Figure 3.6 denotes a unit, and all units are labeled with numerals.

For independent units, bridge-reliability for Figure 3.6 is given by [12]

$$R_b = 2R_1R_2R_3R_4R_5 + R_1R_3R_5 + R_2R_3R_4 + R_2R_5 + R_1R_4 - R_1R_2R_3R_4 - R_1R_2R_3R_5 - R_2R_3R_4R_5 - R_1R_2R_4R_5 - R_3R_4R_5R_1 \quad (3.40)$$

where

$R_b$ is the bridge network's reliability, and

$R_j$ is the unit $j$ reliability, for $j = 1,2,3,4,5$.

For identical units, Equation 3.40 simplifies to

$$R_b = 2R^5 - 5R^4 + 2R^3 + 2R^2 \quad (3.41)$$

where

$R$ is the unit reliability.

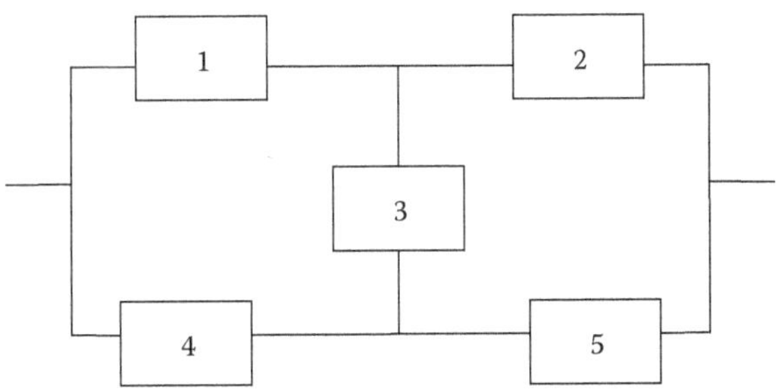


**FIGURE 3.6**
Five nonidentical unit bridge network.

For constant failure rates of all five units, with the aid of Equations 3.11 and 3.41, we obtain:

$$R_b(t) = 2e^{-5\lambda t} - 5e^{-4\lambda t} + 2e^{-3\lambda t} + 2e^{-2\lambda t} \tag{3.42}$$

where
$\lambda$ is the unit constant failure rate.
$R_b(t)$ is the bridge network's reliability at time $t$.

By inserting Equation 3.42 into Equation 3.13, we obtain

$$\begin{aligned} MTTF_b &= \int_0^{\infty} (2e^{-5\lambda t} - 5e^{-4\lambda t} + 2e^{-3\lambda t} + 2e^{-2\lambda t})dt \\ &= \frac{49}{60\lambda} \end{aligned} \tag{3.43}$$

where
$MTTF_b$ is the bridge network's mean time to failure.

**EXAMPLE 3.9**

Assume that in a robot system, five identical and independent units form a bridge network. Calculate the bridge network's reliability for a 250-h mission and mean time to failure, if each unit's constant failure rate is 0.0005 failures per hour.

By inserting the given data values into Equation 3.42, we obtain:

$$\begin{aligned} R_b(250) &= 2e^{-5(0.0005)(250)} - 5e^{-4(0.0005)(250)} + 2e^{-3(0.0005)(250)} + 2e^{-2(0.0005)(250)} \\ &= 0.9700 \end{aligned}$$

Similarly, substituting the specified data value into Equation 3.43 yields:

$$\begin{aligned} MTTF_b &= \frac{49}{60(0.0005)} \\ &= 1633.33 \text{ h} \end{aligned}$$

Thus, the bridge network's reliability and mean time to failure are 0.9700 and 1633.33 h, respectively.

## 3.5 Need for Safety and the Role of Engineers with Respect to Safety

The desire to be safe and secure has always been an important concern to humans, going back to 2300 BC, with the laws of the Sumerian, Akkadian, Babylonian and Abyssinian empires (modern-day Iraq and Syria), including the codes of Urukagina, Ur-Nammu and Hammurabi, among others. The best known is the Code of Hammurabi [5,13].

Today, safety has become a very important issue because each year a vast number of people die and get seriously injured in workplace and other accidents around the globe. For example, in the United States alone in 1996, as per the United States National Safety Council, there were 93,400 deaths and a very large number of disabling injuries due to accidents [10,14]. The cost of these accidents was estimated to be about US$121 billion.

Factors that play a key role in demanding better safety include government regulations, an increasing number of law suits, and public pressure.

Today, engineering products or systems have become very complex and sophisticated. The matter of safety related to such products or systems has become a challenging issue for engineers who are pressured to complete new designs rapidly and at lower costs because of competition and other factors. Past experiences clearly indicate that this, in turn, generally leads to more design-related deficiencies, errors, and causes of accidents. Design-related deficiencies can contribute to or cause accidents.

As per [15], design-related deficiency may result because a designer/design:

- Fails to eliminate or reduce the occurrence of human error
- Offers confusing, wrong, or unfinished concepts/products
- Fails to prescribe an appropriate operational procedure in conditions where a hazard might exist
- Violates normal tendencies/capabilities of potential users
- Fails to properly warn of a potential hazard
- Incorporates weak warning mechanisms rather than providing a safe design for eradicating hazards
- Fails to provide an acceptable level of protection in a user's personal protective equipment
- Relies on users of product/item to avoid an accident
- Fails to foresee an unexpected application of an item/product/system or its potential consequences
- Creates an arrangement of operating controls and other devices that significantly increase reaction time during an emergency or is conducive to errors

- Creates an unsafe characteristic of an item/product/system
- Does not properly determine or consider the consequences of an action, error, failure, or omission
- Places an unreasonable level of stress on operators/users

## 3.6 Classifications of Product Hazards and Common Mechanical Injuries

There are many product-related hazards. As per [16], these may be grouped under the following six classifications:

- *Misuse-and-abuse hazards.* These hazards are concerned with product-usage by humans. Past experiences clearly indicate that product-misuse can lead to serious injuries. Product-abuse can also lead to hazardous situations or injuries; two examples of abuse are poor operating practices and lack of proper maintenance.
- *Electrical hazards.* Two principal components of electrical hazards are electrocution and shock. Past experiences indicate that the major electrical hazard to product/system/property stems from electrical faults, frequently referred to as short circuits.
- *Human-factor hazards.* These hazards are associated with poor design with regard to humans, more specifically, to their physical strength, height, intelligence, computational ability, education, weight, length of reach, visual angle, and so on.
- *Kinematic hazards.* These hazards are associated with situations where parts/items come together while moving and lead to pinching, crushing, or cutting of any object/item caught between them.
- *Environmental hazards.* These hazards may be categorized under two groups: internal hazards and external hazards. The internal hazards are associated with the changes in the surrounding environment that lead to internal damage in the product. This type of hazard can be eliminated or minimized by carefully considering, during the design phase, factors such as extremes of temperatures, vibrations, atmospheric contaminants, ambient noise level, and electromagnetic radiation.

  The external hazards are the hazards posed by the product/system during its life span. These hazards include service-life operational hazards, disposal hazards, and maintenance hazards.
- *Energy hazards.* These hazards may be categorized under two groups: kinetic energy hazards and potential energy hazards. The kinetic energy hazards pertain to items/parts that have energy because of their motion. Three examples of such items are fly wheels, loom

shuttles, and fan blades. Any object that interferes with their motion can experience extensive damage.

The potential energy hazards pertain to items/parts that store energy. Examples of such items/parts include springs, compressed-gas receivers, electronic capacitors, and counterbalancing weights. During the equipment-servicing processes, such hazards are important because stored energy can result in serious injury when released suddenly.

In the industrial sector, humans interact with various types of equipment to carry out tasks such as cutting, chipping, drilling, and stamping. There are various types of injuries that can occur in performing such tasks. Some of the commonly occurring injuries are as follows [5,10]:

- *Breaking injuries.* Generally, such injuries are associated with machines used to deform various types of engineering materials and can result in various types of fractures—simple fracture, oblique fracture, transverse fracture, and complete fracture.
- *Crushing injuries.* Such injuries occur when a body part is caught between two hard surfaces moving progressively together and crushing any object/item that comes between them.
- *Shearing injuries.* Such injuries pertain to shearing processes and include tragic damages such as amputation of fingers and hands.
- *Straining-and-spraining injuries.* Such injuries are generally associated with the use of machines or other tasks and include spraining of ligaments or straining of muscles.
- *Puncturing injuries.* Such injuries occur in a situation when an object penetrates sharply into an individual's body. In an industrial setting, generally, these injuries arise from punching machines because they have sharp tools.
- *Tearing and cutting injuries.* Such injuries occur when an individual's body part comes in contact with a sharp edge.

## 3.7 Organization Tasks for Product Safety and Safety Management Principles

An organization concerned with product safety performs various types of tasks. Some of these tasks are as follows [6,17]:

- Review safety-related customer complaints and field reports
- Review product-test reports to determine shortcomings or trends with regard to safety

- Review government and non-government requirements related to product safety
- Develop directives and programs related to product safety
- Review warning labels—regarding safety factors such as meeting requirements of, adequacy, and compatibility to warnings in the instruction manuals—that are to be placed on products
- Develop appropriate mechanisms by which the safety program can be monitored effectively
- Review mishaps and hazards in existing similar products/items for avoiding repetition of such hazards in new products/items
- Review the product to establish whether the potential hazards have been eradicated or controlled
- Determine if items such as emergency equipment, warning and monitoring devices, or protective equipment are required in handling or using the product
- Develop appropriate safety criteria based on applicable voluntary and governmental standards for use by the vendor, subcontractor, and company design professionals
- Provide appropriate assistance to designers in choosing alternative means to eradicate or control hazards or other safety-related problems in initial designs
- Review proposed product maintenance and operation documents with regard to safety
- Participate in reviewing accident-associated claims or recall actions by government agencies and recommend appropriate remedial actions for justifiable recalls or claims

Over the years, professionals working in the area of safety have developed various safety management principles. Some of these principles are as follows [6,18,19]:

- The function of safety is to find out and define the operational errors that result in accidents.
- The key to successful line-safety performance is management procedures that clearly factor in accountability.
- The causes that lead to unsafe behavior can be identified, controlled, and classified; the classifications of the causes include overload, the employee's decision to err, and traps.
- Safety should be managed just like any other function in an organization/company—more clearly, management should direct safety by setting goals that can be attained, and by planning, organizing, and controlling the attainment of such goals successfully.

- There are specific sets of circumstances that can predictably lead to severe injuries—nonproductive activities, certain construction conditions, high-energy sources, and abnormal nonroutine activities.
- The symptoms that indicate something is incorrect in the management system include accidents, unsafe actions, and unsafe conditions.
- Under most circumstances, normal human behavior is an unsafe behavior. Thus, it is the management team's responsibility as leaders to make appropriate changes to the environment that fosters unsafe behavior in order to encourage safe behavior.
- In developing a good safety system, the three main subsystems that must be considered with care are the behavioral, the managerial, and the physical subsystems.
- There is no single method to effectively achieve safety in an organization. But, for a safety system to be effective, it must meet certain criteria, such as: have the top management visibly showing its full support to safety, involve worker participation, be flexible, and so on.

## 3.8 Accident-Causation Theories

As per [5], there are many accident-causation theories. Two such theories are presented below.

### 3.8.1 The "Human Factors" Accident-Causation Theory

The "human factors" accident-causation theory is based on the assumption that accidents occur due to a chain of events directly or indirectly linked to human error. It consists of three main factors, shown in Figure 3.7, that lead to the occurrence of human error [5,20]. These three main factors are: inappropriate response/incompatibility; overload; and inappropriate activities.

The factor "inappropriate response/incompatibility" is an important factor for the occurrence of human error and some examples of inappropriate response by an individual are as follows [6,20]:

- A person completely disregards the specified safety procedures
- A person removes a safeguard from a machine for increasing output
- A person detects a hazardous condition, but takes no corrective action

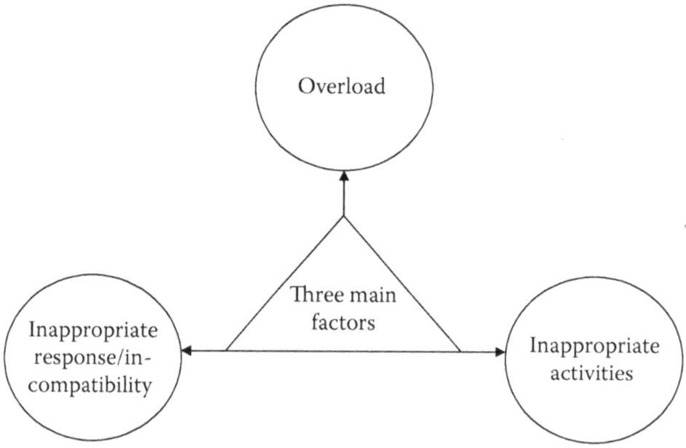


**FIGURE 3.7**
Three main factors that lead to the occurrence of human error.

The "overload" factor is concerned with an imbalance between the capacity of an individual at any time and the load he/she is carrying in a given state. The capacity of a person is the product of factors such as:

- The degree of training
- Fatigue
- Physical condition
- Stress
- State of mind
- Natural ability

The load carried by an individual is made up of tasks for which he/she has responsibility, along with additional burdens resulting from the situational factors (i.e., level of risk, unclear instructions, etc.), environmental factors (i.e., distractions, noise, etc.), and internal factors (i.e., personal problems, worry, emotional stress, etc.).

Finally, the factor of "inappropriate activities" is concerned with inappropriate activities undertaken due to human error, poor judgment about the degree of risk involved in a given task and subsequently acting on that misjudgment.

### 3.8.2 The "Domino" Accident-Causation Theory

This theory is encapsulated in 10 statements called the "Axioms of Industrial Safety" by H.W. Heinrick [21]. These 10 axioms are as follows [5,6,21]:

i. Most accidents are due to the unsafe acts of people.
ii. An unsafe act by a person or an unsafe condition does not always immediately lead to an accident/injury.

iii. Supervisors play a key role in the prevention of industrial accidents.
iv. The reasons why people commit unsafe acts can be useful in choosing appropriate corrective measures.
v. The severity of an injury is largely fortuitous, and the specific accident that caused it is generally preventable.
vi. There are two types of costs (i.e., direct and indirect) of an accident. Some examples of the direct costs are liability claims, hospital-related expenses, and compensation.
vii. The occurrence of injuries results from a completed sequence of numerous factors, the last one of which is the accident itself.
viii. Management should assume full responsibility for safety because it is in the best position for achieving the final results effectively.
ix. An accident can occur only when someone commits an unsafe act and/or there is a mechanical or physical hazard.
x. The most useful accident-prevention methods are analogous to the productivity and quality approaches.

According to Heinrich, the five specific factors in the sequence of events leading up to an accident are as follows [5,6]:

- *Ancestry and social environment.* In this factor, it is assumed that negative character traits such as recklessness, stubbornness, and avariciousness that might lead individuals to behave in an unsafe manner, can be acquired as a result of the social environment or surroundings or inherited through one's ancestry.
- *Fault of a person.* In this factor, it is assumed that negative character traits (whether inherited or acquired) such as ignorance of safety practices, excitability, recklessness, nervousness, and violent temper constitute proximate reasons for committing unsafe acts or for the existence of physical or mechanical hazards.
- *Unsafe act/physical or mechanical hazard.* In this factor, it is assumed that unsafe acts by people (i.e., starting equipment/machinery without warning, removing safeguards, standing under suspended loads) and physical or mechanical hazards (i.e., unguarded point of operation, unguarded gears, inadequate light, absence of guardrails) are the very direct causes for the occurrences of accidents.
- *Accident.* In this factor, it is assumed that events such as falls of humans and striking of humans by flying objects are the typical examples of accidents that cause injury.
- *Injury.* In this factor, it is assumed that the typical injuries that directly result from accidents include fractures and lacerations.

## 3.9 Problems

1. What are the three phases of a bathtub hazard rate curve? Discuss the causes of failures in each of these three phases.
2. Write down general equations for the following:
   i. Reliability function
   ii. Hazard-rate function
3. Prove Equation 3.11 by using Equation 3.6.
4. Write down formulas to obtain mean time to failure by using the reliability function.
5. Assume that a robot system is composed of three independent and identical subsystems and the constant failure rate of each subsystem is 0.0005 failures per hour. All three subsystems must operate normally for the robot system to work successfully. Calculate the robot system reliability for a 50-hour mission, mean time to failure, and failure rate.
6. A robot system has four active, identical, and independent units in parallel. At least three units must operate normally for the successful operation of the robot system. Calculate the robot system's mean time to failure if the unit's constant failure rate is 0.006 failures per hour.
7. Prove Equation 3.31 by using Equation 3.35.
8. Discuss the "human factors" accident-causation theory.
9. List at least 10 tasks of a product-related safety organization.
10. Discuss product-hazard classifications.

## References

1. Layman, W.J., Fundamental consideration in preparing a master plan, *Electric World*, Vol. 101, 1933, pp. 778–792.
2. Smith, S.A., Service reliability measured by probabilities of outage, *Electrical World*, Vol. 103, 1934, pp. 371–374.
3. Dhillon, B.S., *Power System Reliability, Safety, and Management*, Ann Arbor Science Publishers, Ann Arbor, Michigan, 1983.
4. Dhillon, B.S., *Design Reliability: Fundamentals and Applications*, CRC Press, Boca Raton, Florida, 1999.
5. Goetsch, D.L., *Occupational Safety and Health*, Prentice-Hall, Englewood Cliffs, New Jersey, 1996.

6. Dhillon, B.S., *Engineering Safety: Fundamentals, Techniques, and Applications*, World Scientific Publishing, River Edge, New Jersey, 2003.
7. Kapur, K.C., Reliability and maintainability, in *Handbook of Industrial Engineering*, Eds. G. Salvendy, John Wiley and Sons, New York, 1982, pp. 8.5.1–8.5.34.
8. Dhillon, B.S., Life distributions, *IEEE Transactions on Reliability*, Vol. 30, No. 5, 1981, pp. 457–460.
9. Shooman, M.L., *Probabilistic Reliability: An Engineering Approach*, McGraw-Hill Book Company, New York, 1968.
10. Dhillon, B.S., *Reliability, Quality, and Safety for Engineers*, CRC Press, Boca Raton, Florida, 2005.
11. Sandler, G.H., *System Reliability Engineering*, Prentice-Hall, Englewood Cliffs, New Jersey, 1963.
12. Lipp, J.P., Topology of switching elements versus reliability, *Transactions on IRE Reliability and Quality Control*, Vol. 7, 1957, pp. 21–34.
13. Ladon, J., Ed., *Introduction to Occupational Health and Safety*, National Safety Council (NSC), Chicago, Illinois, 1986.
14. National Safety Council, *Accidental Facts, Report*, Chicago, Illinois, 1996.
15. Hammer, W., Price, D., *Occupational Safety Management and Engineering*, Prentice-Hall, Upper Saddle River, New Jersey, 2001.
16. Hunter, T.A., *Engineering Design for Safety*, McGraw-Hill, New York, 1992.
17. Hammer, W., *Product Safety Management and Engineering*, Prentice-Hall, Englewood Cliffs, New Jersey, 1980.
18. Petersen, D., *Safety Management*, American Society of Safety Engineers, Des Plaines, Illinois, 1998.
19. Petersen, D., *Techniques of Safety Management*, McGraw-Hill, New York, 1971.
20. Heinrich, H.W., Petersen, D., Roos, N., *Industrial Accident Prevention*, McGraw-Hill, New York, 1980.
21. Heinrich, H.W., *Industrial Accident Prevention*, McGraw-Hill, New York, 1959.

# 4

# *Methods for Performing Reliability and Safety Analysis of Robot Systems*

## 4.1 Introduction

Over the years, a large number of publications in the areas of reliability and safety have appeared in the form of articles in journals, related to conference proceedings, technical reports, and books [1–6]. Many of these publications report the development of various types of methods and approaches for analyses of reliability and safety. Some of these methods and approaches can be used quite effectively for analyzing both reliability and safety. The others are more confined to a specific area (i.e., reliability or safety).

Three examples of these methods and approaches that can be used to analyze both reliability and safety are failure modes and effect analysis (FMEA), the Markov method, and fault-tree analysis (FTA). FMEA was developed in the early 1950s to analyze engineering systems from the aspect of reliability. The Markov method is named after a Russian mathematician, Andrei A. Markov (1856–1922), and is a highly mathematical method that is frequently used for performing reliability and safety analyses in engineering systems. Finally, the FTA method was developed in the early 1960s to analyze rocket-launch control systems from the safety aspect.

Today, these three methods are used across many diverse areas to analyze various types of problems related, directly or indirectly, to reliability and safety [5–9]. This chapter presents a number of methods and approaches considered useful in analyzing the reliability or safety or both of robot systems. All have been extracted from the published literature in the areas of reliability and safety.

## 4.2 Failure Modes and Effect Analysis

FMEA is a widely used method for analyzing engineering systems for their reliability. The method may simply be described as an effective approach for analyzing each potential failure mode in the system to examine the effects of such failure modes on the entire system [10].

The history of FMEA may be traced back to early 1950s with the development of flight control systems, when the Bureau of Aeronautics of the U.S. Navy, in order to develop a procedure for controlling reliability in the detail design effort, developed a requirement known as Failure Analysis [11]. Subsequently, the term "Failure Analysis" was changed to FMEA. In the 1970s, the U.S. Department of Defense directed its effort to developing a military standard entitled "Procedures for Performing a Failure Mode, Effects, and Criticality Analysis (FMECA)" [12]. FMECA is an extended version of FMEA. More clearly, when FMEA is extended to group each potential failure effect with respect to its severity (this incorporates documenting catastrophic and critical failures), the method is called FMECA [5,6,13].

The seven main steps followed to perform FMEA are as follows [2,5]:

- *Step 1*: Define system boundaries and its associated detailed requirements.
- *Step 2*: List system subsystems and parts/components.
- *Step 3*: List each part's/component's failure modes, the description and the identification.
- *Step 4*: Assign the failure-occurrence probability/rate to each part's/component's failure mode.
- *Step 5*: List each failure mode's effect/effects on the subsystem, system, and plant.
- *Step 6*: Enter remarks for each failure mode.
- *Step 7*: Review each critical failure mode and take necessary action.

It is to be noted that there are many factors that must be explored with care prior to the implementation of FMEA. These factors include examination of each conceivable failure mode by the involved professionals, measuring FMEA cost/benefits, making decisions based on the risk priority number (RPN), and obtaining the engineer's approval and support [14].

Over the years, professionals involved with reliability-analysis have established certain guidelines/facts concerning FMEA. Five of these guidelines/facts are shown in Figure 4.1 [14]. Nonetheless, there are many benefits to performing FMEA. Some of the main ones are as follows [5,14]:

- Easy to understand
- A useful tool for improving communication among design interface personnel
- A systematic approach for classifying hardware failures
- Reduces development time and cost

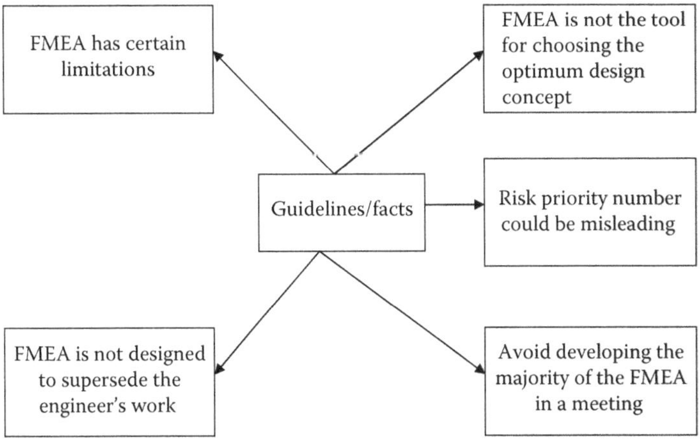


**FIGURE 4.1**
FMEA-related guidelines/facts.

- A useful approach that starts from the detailed level and works upward
- Identifies safety-related concerns for focus
- A visibility tool for management that also improves customer satisfaction
- A useful approach for comparing designs; reduces engineering changes
- A useful approach for safeguarding against repeating the same mistakes in the future
- A useful approach for more efficient test planning

Additional information on this method is available in [5,14].

## 4.3 The Markov Method

The Markov method is widely used to perform various types of reliability-analysis of engineering systems and is named after a Russian mathematician, Andrei A. Markov (1856–1922). The method is used frequently to model reparable systems with constant failure and repair rates.

The method is subject to the following assumptions [5,15]:

- All occurrences are independent of each other.

- The transitional probability from one system-state to another in the finite time interval $\Delta t$ is given by $\alpha\Delta t$, where $\alpha$ is the transition rate (e.g., failure or repair rate) from one system state to another.
- The probability of more than one transition occurrence in the finite time interval $\Delta t$ from one system state to another is negligible (e.g., $[\alpha\Delta t][\alpha\Delta t] \rightarrow 0$).

The following example demonstrates the application of this method.

**EXAMPLE 4.1**

Assume that a robot system can either be in an operating or a failed state and its constant failure and repair rates are $\alpha_r$ and $\mu_r$, respectively. The robot-system state-space diagram is shown in Figure 4.2. The numerals in the box and circle denote the robot-system states. Develop expressions for the robot system, with the aid of the Markov method, time-dependent and steady-state availabilities and unavailabilities, reliability, and mean time to failure.

With the aid of the Markov method, we write down the following equations for states 0 and 1, respectively shown in Figure 4.2:

$$P_0(t + \Delta t) = P_0(t)(1 - \alpha_r\Delta t) + P_1(t)\mu_r\Delta t \tag{4.1}$$

$$P_1(t + \Delta t) = P_1(t)(1 - \mu_r\Delta t) + P_0(t)\alpha_r\Delta t \tag{4.2}$$

where

$t$ is time,

$\alpha_r\Delta t$ is the probability of robot system failure in finite time interval $\Delta t$,

$(1 - \alpha_r\Delta t)$ is the probability of no robot system failure in finite time interval $\Delta t$,

$\mu_r\Delta t$ is the probability of robot system repair in finite time interval $\Delta t$,

$(1 - \mu_r\Delta t)$ is the probability of no robot system repair in finite time interval $\Delta t$,

$P_j(t)$ is the probability that the robot system is in state $j$ at time $t$, for $j = 0,1$,

$P_0(t + \Delta t)$ is the probability of the robot system being in operating state 0 at time $(t + \Delta t)$, and

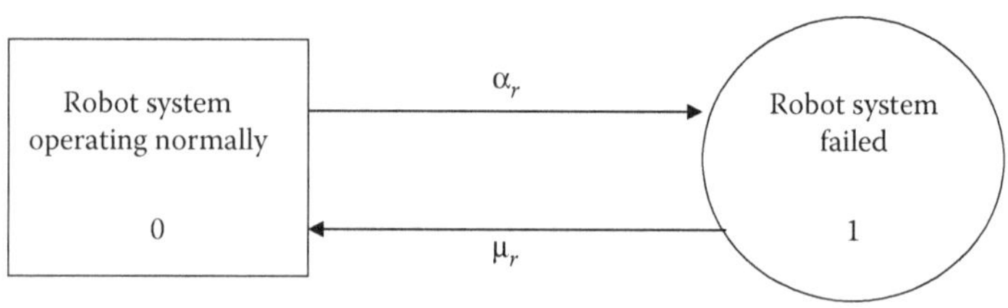


**FIGURE 4.2**
Robot system state-space diagram.

$P_1(t + \Delta t)$ is the probability of the robot system being in failed state 1 at time $(t + \Delta t)$.

From Equation 4.1, we write

$$P_0(t + \Delta t) = P_0(t) - P_0(t)\alpha_r \Delta t + P_1(t)\mu_r \Delta t \tag{4.3}$$

From Equation 4.3, we get

$$\lim_{\Delta t \to 0} \frac{P_0(t + \Delta t) - P_0(t)}{\Delta t} = -P_0(t)\alpha_r + P_1(t)\mu_r \tag{4.4}$$

Thus, from Equation 4.4, we obtain

$$\frac{dP_0(t)}{dt} + P_0(t)\alpha_r = P_1(t)\mu_r \tag{4.5}$$

Similarly, using Equation 4.2, we obtain

$$\frac{dP_1(t)}{dt} + P_1(t)\mu_r = P_0(t)\alpha_r \tag{4.6}$$

At time $t = 0$, $P_0(0) = 1$, and $P_1(0) = 0$.
By solving Equations 4.5 and 4.6, we get [5]

$$P_0(t) = \frac{\mu_r}{(\alpha_r + \mu_r)} + \frac{\alpha_r}{(\alpha_r + \mu_r)} e^{-(\alpha_r + \mu_r)t} \tag{4.7}$$

$$P_1(t) = \frac{\alpha_r}{(\alpha_r + \mu_r)} - \frac{\alpha_r}{(\alpha_r + \mu_r)} e^{-(\alpha_r + \mu_r)t} \tag{4.8}$$

Thus, the robot system time-dependent availability and unavailability, respectively, are

$$A_r(t) = P_0(t) = \frac{\mu_r}{(\alpha_r + \mu_r)} + \frac{\alpha_r}{(\alpha_r + \mu_r)} e^{-(\alpha_r + \mu_r)t} \tag{4.9}$$

and

$$UA_r(t) = P_1(t) = \frac{\alpha_r}{(\alpha_r + \mu_r)} - \frac{\alpha_r}{(\alpha_r + \mu_r)} e^{-(\alpha_r + \mu_r)t} \tag{4.10}$$

where
$A_r(t)$ is the robot system availability at time $t$ and
$UA_r(t)$ is the robot system unavailability at time $t$.

By letting time $t$ go to infinity in Equations 4.9 and 4.10, we obtain [5]:

$$A_r = \lim_{t \to \infty} A_r(t) = \frac{\mu_r}{\alpha_r + \mu_r} \quad (4.11)$$

and

$$UA_r = \lim_{t \to \infty} UA_r(t) = \frac{\alpha_r}{\alpha_r + \mu_r} \quad (4.12)$$

where
$A_r$ is the robot system steady-state availability and
$UA_r$ is the robot system steady-state unavailability.

For $\mu_r = 0$, from Equation 4.9 we obtain

$$R_r(t) = P_0(t) = e^{-\alpha_r t} \quad (4.13)$$

where
$R_r(t)$ is the robot system reliability at time $t$.

By integrating Equation 4.13 over the time interval $[0,\infty]$, we obtain the following equation for the robot-system mean time to failure [5]:

$$MTTF_r = \int_0^{\infty} e^{-\alpha_r t} dt = \frac{1}{\alpha_r} \quad (4.14)$$

where
$MTTF_r$ is the robot system mean time to failure.

Thus, the robot-system time-dependent and steady-state availabilities and unavailabilities, reliability, and mean time to failure are given by Equations 4.9 through 4.14.

**EXAMPLE 4.2**

Assume that a robot system's constant failure and repair rates are 0.0006 failures per hour, and 0.005 repairs per hour, respectively. Calculate the robot system steady-state unavailability and unavailability during a 20 h mission.

By inserting the specified data values into Equations 4.12 and 4.10, we get

$$UA_r = \frac{0.0006}{0.0006 + 0.005} = 0.1071$$

and

$$UA(20) = \frac{0.0006}{(0.0006 + 0.005)} - \frac{0.0006}{(0.0006 + 0.005)} e^{-(0.0006+0.005)(20)}$$
$$= 0.013$$

Thus, the robot-system steady-state unavailability and unavailability during a 20 h mission are 0.1071 and 0.113, respectively.

## 4.4 Fault Tree Analysis

This method was developed in the early 1960s at the Bell Telephone Laboratories for performing reliability-analysis of the Minuteman Launch Control System [2,5]. Today, it is widely used to evaluate the reliability of engineering systems in their design and development, particularly in the area of nuclear power generation.

A fault tree may simply be described as a logical representation of the relationship of primary fault events that lead to the occurrence of a stated undesirable event called the "top event" and is depicted using a tree structure with logic gates such as OR and AND. Some of the purposes of performing FTA are as follows [5]:

- Understanding the level of protection that the design concept provides against failures
- Understanding the functional relationship of system failures
- Meeting jurisdictional requirements
- Identifying critical areas and potential for cost-effective improvements

FTA begins by identifying an undesirable event, called the "top event," associated with a system under consideration. Fault events that can lead to the occurrence of the top event are generated and connected by logic operators such as OR and AND. The OR gate provides a true output (i.e., fault) if one or more inputs (i.e., faults) are true, whereas the AND gate provides a true output (i.e., fault) if all the inputs (i.e., faults) are true.

The construction of a fault tree proceeds by generating fault events in a successive manner until the fault events need not be developed any further. These fault events are known as primary or basic events. During the process of fault tree construction, one successively asks the question: "How could

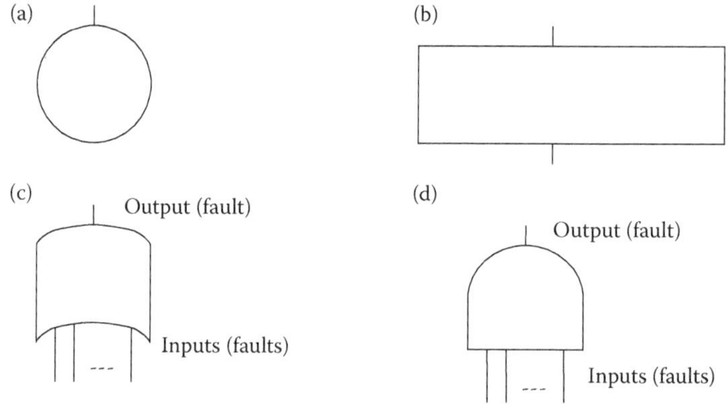


**FIGURE 4.3**
Fault tree symbols: (a) circle, (b) rectangle, (c) OR gate, (d) AND gate.

this fault event occur?" Basic symbols used in constructing fault trees are shown in Figure 4.3 [2,5].

All the four symbols shown in Figure 4.3 are described below (it is to be noted that for the sake of clarity OR and AND gates are described again).

- *Circle*: It denotes a basic or primary fault event (e.g., failure of an elementary part or component). The parameters of the event are probability of occurrence, failure rate, and repair rate; whose values are normally obtained from empirical data.
- *Rectangle*: It denotes a fault event that occurs from the logical combination of fault events through the input of a logic gate such as AND and OR.
- *OR gate*: It represents an output fault event that occurs if one or more of the input fault events occur.
- *AND gate*: It represents an output event that occurs only if all of the input fault events occur.

Information on other symbols used in conducting FTA is available in [2,5].

**EXAMPLE 4.3**

Assume that a windowless room containing a robot system has three light bulbs (i.e., *X*,*Y*,*Z*) and one switch. The switch can only fail to close. Develop a fault tree for the undesirable event (i.e., top fault event) "dark robot system room" (i.e., no light in the room containing the robot system) using symbols presented in Figure 4.3.

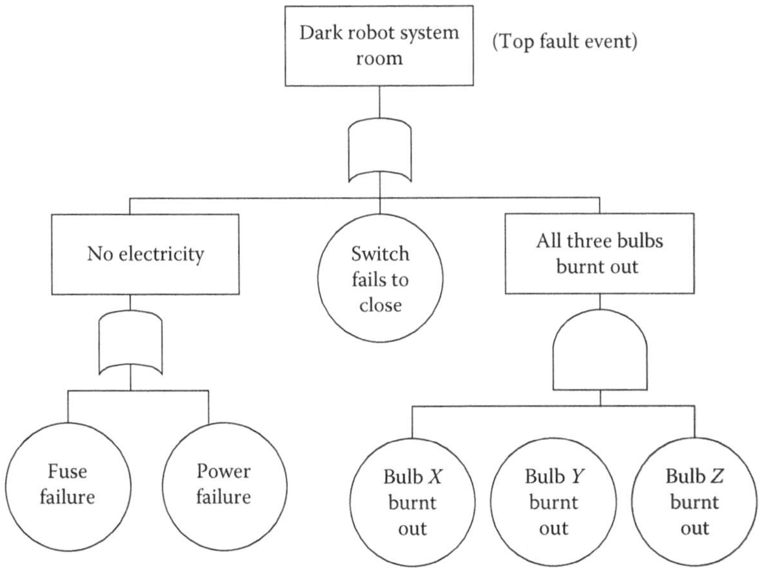


**FIGURE 4.4**
A fault tree for the top fault event: dark robot system room.

In this case, the robot system room can only be dark if the switch fails to close, if all the three light bulbs burn out, or if there is no incoming electricity. A fault tree for the example is shown in Figure 4.4.

### 4.4.1 Probability Evaluation of Fault Trees

Under certain circumstances, it may be necessary to predict the occurrence probability of a certain event (e.g., no light in the room containing the robot system). Before this could be achieved by using the FTA approach, the calculation of the probability of occurrence of the output-fault events of logic gates is necessary.

Thus, the probability of occurrence of the output-fault event of an AND gate is expressed by [2,5]

$$P_a(Y) = \prod_{j=1}^{m} P(Y_j) \tag{4.15}$$

where

$P_a(Y)$ is the probability of occurrence of AND gate's output-fault event $Y$,
$m$ is the number of independent input-fault events and
$P(Y_j)$ is the probability of occurrence of input-fault event $Y_j$, for $j = 1,2,3,\ldots,m$.

Similarly, the probability of occurrence of the output-fault event of an OR gate is given by [2,5]

$$P_o(Y) = 1 - \prod_{j=1}^{m}(1 - P(Y_j)) \tag{4.16}$$

where

$P_0(Y)$ is the probability of occurrence of OR gate's output-fault event $Y$.

**EXAMPLE 4.4**

Assume that in Figure 4.4 the probabilities of occurrence of fault events—power failure, fuse failure, switch fails to close, burnt-out bulb $X$, burnt-out bulb $Y$, and burnt-out bulb $Z$—are 0.02, 0.03, 0.04, 0.05, 0.05, and 0.05, respectively. With the aid of Equations 4.15 and 4.16, calculate the probability of occurrence of the top fault event "dark robot system room" and redraw Figure 4.4 with the given and calculated probability values.

By substituting the given data values into Equation 4.15, the probability of occurrence of the event "all three bulbs burnt out" is

$$\begin{aligned} P_b &= (0.05)(0.05)(0.05) \\ &= 0.000125 \end{aligned}$$

where

$P_b$ is the probability of occurrence of the event "all three bulbs burnt out."

Similarly, by inserting the specified data values into Equation 4.16, the probability of occurrence of the event "no electricity" is

$$\begin{aligned} P_n &= (0.03) + (0.02) - (0.03)(0.02) \\ &= 0.0494 \end{aligned}$$

where

$P_n$ is the probability of occurrence of the event "no electricity."

By inserting the above calculated values and the specified data value into Equation 4.16, the probability of occurrence of the top fault event "dark robot system room" is

$$\begin{aligned} P_d &= 1 - (1 - 0.0494)(1 - 0.04)(1 - 0.000125) \\ &= 0.0875 \end{aligned}$$

Thus, the probability of occurrence of the top fault event "dark robot system room" is 0.0875. Figure 4.4 with the given data and calculated values is shown in Figure 4.5.

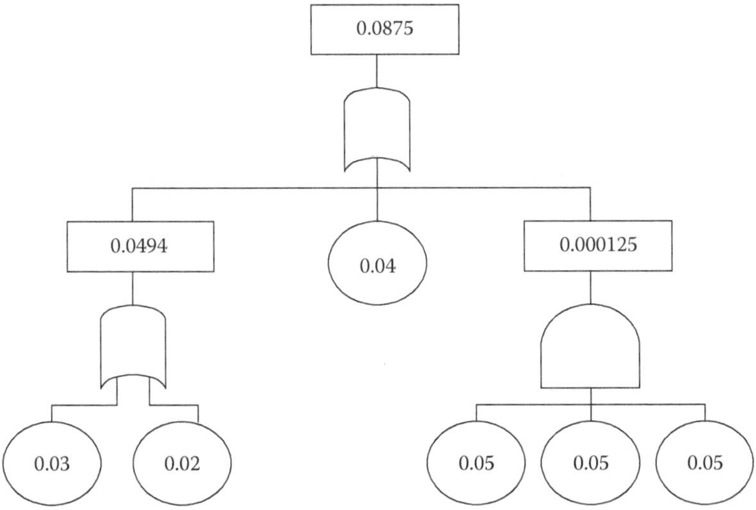


**FIGURE 4.5**
A fault tree with the given and calculated event occurrence probability values.

### 4.4.2 FTA Advantages and Disadvantages

There are many advantages and disadvantages of performing FTA. Some of its advantages are as follows [2,5]:

- Allows concentration on one specific failure at a time
- Offers a graphic aid for management
- Useful to handle complex systems more easily
- Requires the analyst to understand thoroughly the system under consideration before starting the analysis
- Useful to provide options for management and others to perform either qualitative or quantitative analysis
- Useful to highlight failures deductively and to provide insight into the system behavior

In contrast, some of the disadvantages of the FTA are as follows [2,5]:

- A time-consuming and costly method
- The end results are difficult to check
- It considers parts or components in either working or failed state (i.e., the partial failure states of the parts or components are difficult to handle)

Additional information on this method is available in [2,5].

## 4.5 Technique of Operations Review

Technique of operations review (TOR) may simply be described as a hands-on analytical approach to identify the root system causes of an operation failure; it was developed in the early 1970s by D.A. Weaver of the American Society of Safety Engineers [6,16]. The method seeks to highlight systemic causes for an adverse incident rather than assigning blame with respect to safety. Thus, it permits management personnel and workers to work jointly to analyze workplace-related accidents, failures, and incidents.

The method is activated by an adverse incident occurring at a certain point in time and location involving certain individuals, and makes use of a worksheet containing simple terms that require yes/no responses. It is to be noted that the method is not a hypothetical process and it demands a systematic evaluation of the circumstances surrounding the incident in question. Ultimately, TOR highlights how the company/organization could have prevented the occurrence of the accident.

The following steps are associated with the method [6,17]:

- *Step 1*: Form the TOR team with appropriate members belonging to all concerned areas.
- *Step 2*: Hold a roundtable session to impart common knowledge to all members of the TOR team.
- *Step 3*: Identify one important systemic factor that played a key role in the occurrence of the accident/incident. It is to be noted that this factor must be based on the consensus of all members of the TOR team and it serves as a starting point for further investigation to follow.
- *Step 4*: Make use of the team consensus in responding to a sequence of yes/no options.
- *Step 5*: Evaluate the identified factors, ensuring that there is clear-cut consensus among the team personnel with respect to each factor.
- *Step 6*: Prioritize the contributory factors by starting with the one considered the most serious.
- *Step 7*: Develop necessary preventive/corrective strategies with respect to each and every contributory factor.
- *Step 8*: Implement the strategies.

Finally, it is to be noted that the main strength of this method is the involvement of line personnel in the analysis and its main weakness is that it is an after-the-fact process.

## 4.6 Hazard and Operability Analysis

This method is considered a powerful tool to identify safety-related problems prior to the availability of complete data concerning an item and it was developed for application in the chemical industry sector [18,19]. Three basic objectives of Hazard and Operability (HAZOP) analysis are as follows [17,20,21]:

- To review each part of the facility/process for determining how deviations from the intended design can happen
- To decide whether the deviations can lead to operating problems/hazards
- To produce a complete description of the facility/process

The seven steps followed to perform HAZOP analysis are shown in Figure 4.6 [17,19,22].

Finally, it is added that this method has basically the same weaknesses as FMEA discussed earlier in the chapter. For example, both HAZOP analysis and FMEA predict problems that are connected to process/system faults/failures, but overlook the factoring in of human error into the equation. This is an important weakness because, frequently, human error is a factor in accident occurrences.

## 4.7 Interface Safety Analysis

This method is concerned with determining the incompatibilities between subsystems and assemblies of an equipment/product that could lead to accidents. The method establishes that distinct units/parts can be integrated into a viable system and that normal functioning of an individual unit/part will not impair the performance of or damage another unit/part or the complete equipment/system. Although the method considers various relationships, they can be grouped under the following three classifications [6,23]:

- *Classification I: Functional relationships.* These relationships are concerned with multiple items or units. For example, in situations where outputs of an item or unit constitute the inputs to the downstream item(s) or unit(s), any error in outputs and inputs may result in damage to the downstream item(s) or unit(s), thereby creating a safety problem or hazard. Such outputs could be in conditions such as erratic outputs, degraded outputs, excessive outputs, unprogrammed outputs, and zero outputs.

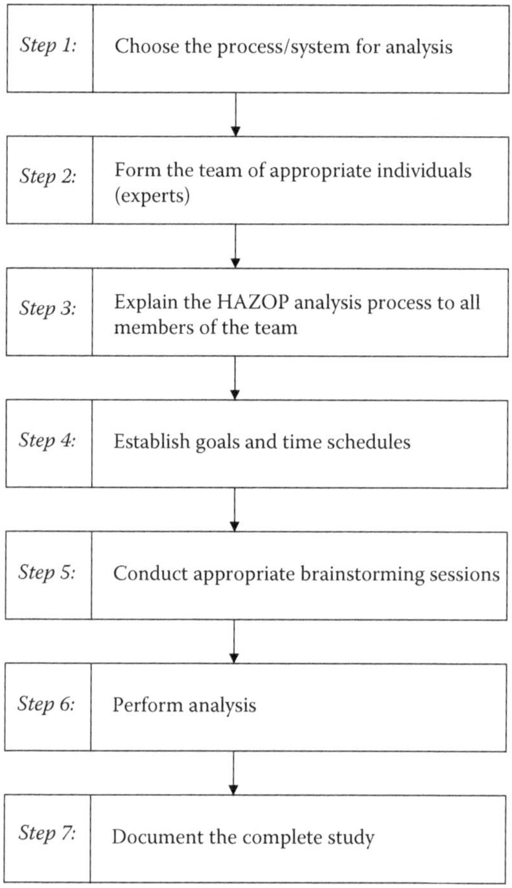


**FIGURE 4.6**
HAZOP analysis steps.

- *Classification II: Flow relationships.* These relationships are concerned with two or more items or units. For example, the flow between two items may involve electrical energy, steam, water, air, lubricating oil, or fuel. In addition, the flow could also be unconfined, such as heat radiation from one item to another.

  Often, the problems experienced with many products or items include the proper flow of fluids and energy from one unit to another unit through confined passages, consequently leading to safety problems. The causes of flow problems include total or partial interconnection failure and faulty connections between units. In the case of fluids, from the safety perspective, the factors that must be considered with utmost care include contamination, flammability, lubricity, loss of pressure, odor, and toxicity.

- *Classification III: Physical relationships.* These are concerned with the physical aspects of products/items. For example, two products or items might be well designed and manufactured and operate quite well individually, but they may have problems in fitting together because of dimension differences, or there may be other incompatibilities that may result in safety problems. Some examples of the other problems are as follows:
  - A small clearance between units; thus, during the removal process, the units may be damaged
  - Situation where it is impossible to tighten, join, or mate parts/units properly
  - Situation where access to or egress from system/equipment is restricted or impossible

## 4.8 Probability Tree Method

The probability tree method is used for performing task analysis by diagrammatically representing critical human actions and other events concerning the system under consideration. Often, the method is used to perform task analysis in the technique for human error rate prediction (THERP) [2,8]. In this method, diagrammatic task analysis is denoted by the probability tree branches. More clearly, the tree's branching limbs denote the outcome (i.e., success or failure) of each event and each branch is assigned a value for the probability of occurrence.

Some of the benefits of the probability-tree method are as follows [2,24]:

- Decreases the probability of errors in computation because of simplification in the computational process
- Includes, with some modifications, factors such as interaction stress, interaction effects, and emotional stress
- A useful visibility tool

The application of this method is demonstrated through the following two examples:

**EXAMPLE 4.5**

Assume that a robot system maintenance person performs two independent safety-related tasks: $x$ and $y$. Each of these two tasks can be performed either correctly or incorrectly and task $x$ is carried out before task $y$.

Develop a probability tree and obtain an expression for the probability of not successfully accomplishing the overall mission by the robot system maintenance person.

In this case, the robot system maintenance person performs task $x$ correctly or incorrectly and then proceeds to perform task $y$. Task $y$ can also be performed correctly or incorrectly by the robot system maintenance person. Figure 4.7 shows this entire scenario.

The symbols used in Figure 4.7 are defined below.

$x$ is the task $x$ performed correctly.
$\bar{x}$ is the task $x$ performed incorrectly.
$y$ is the task $y$ performed correctly.
$\bar{y}$ is the task $y$ performed incorrectly.

By examining Figure 4.7, it is concluded that there is a total of three possibilities (i.e., $x\bar{y}$, $\bar{x}y$, and $\bar{x}\bar{y}$) for not successfully accomplishing the overall mission by the robot system maintenance person. Thus, the probability of not successfully accomplishing the overall mission by the robot system maintenance person is given by

$$\begin{aligned} P_f &= P(x\bar{y}) + P(\bar{x}y) + P(\bar{x}\bar{y}) \\ &= P_x P_{\bar{y}} + P_{\bar{x}} P_y + P_{\bar{x}} P_{\bar{y}} \end{aligned} \tag{4.17}$$

where

$P_f$ is the probability of not successfully accomplishing the overall mission by the robot system maintenance person.
$P_x$ is the probability of performing the independent task $x$ correctly by the robot system maintenance person.
$P_y$ is the probability of performing the independent task $y$ correctly by the robot system maintenance person.
$P_{\bar{x}}$ is the probability of performing the independent task $x$ incorrectly by the robot system maintenance person.

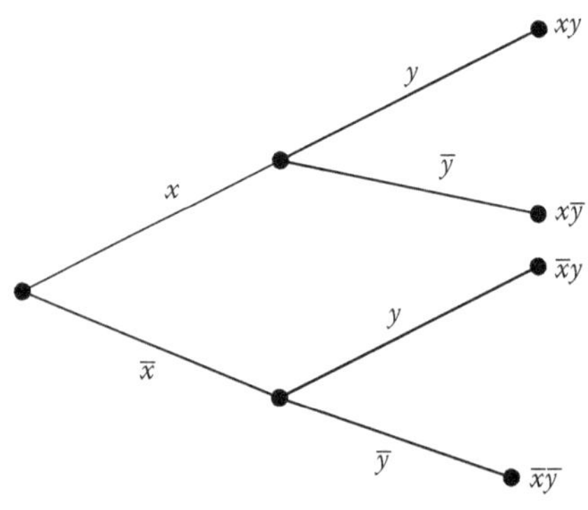


**FIGURE 4.7**
Probability tree diagram for the robot system maintenance person performing tasks $x$ and $y$.

$P_{\bar{y}}$ is the probability of performing the independent task $y$ incorrectly by the robot system maintenance person.

Thus, Equation 4.17 is the expression for the probability of not successfully accomplishing the overall mission by the robot system maintenance person.

**EXAMPLE 4.6**

With the aid of Figure 4.7, for Example 4.5, develop a probability expression for successfully accomplishing the overall mission by the robot system maintenance person. In addition, calculate the probability of not successfully accomplishing the overall mission by the robot system maintenance person, if the probabilities of performing tasks $x$ and $y$ correctly are 0.85 and 0.95, respectively.

Thus, with the aid of Figure 4.7, the probability of successfully accomplishing the overall mission by the robot system maintenance person is

$$\begin{aligned} P_s &= P(xy) \\ &= P_x P_y \end{aligned} \tag{4.18}$$

where

$P_s$ is the probability of successfully accomplishing the overall mission by the robot system maintenance person.

Because the $P_{\bar{x}} = 1 - P_x$ and $P_{\bar{y}} = 1 - P_y$, Equation 4.17 reduces to

$$P_f = 1 - P_x P_y \tag{4.19}$$

By inserting the specified data values into Equation 4.19, we get

$$\begin{aligned} P_f &= 1 - (0.85)(0.95) \\ &= 0.1925 \end{aligned}$$

Thus, the probability of not successfully accomplishing the overall mission by the robot system maintenance person is 0.1925.

## 4.9 Problems

1. What are the main advantages of FMEA?
2. What is the difference between FMECA and FMEA? Describe FMEA.
3. Prove Equations 4.7 and 4.8 by using Equations 4.5 and 4.6.
4. Assume that a robot system's constant failure and repair rates are 0.0004 failures per hour and 0.002 repairs per hour, respectively. Calculate the robot system steady-state unavailability and availability during a 15 h mission.

5. What are the purposes of performing FTA?
6. Describe the following four items used in performing FTA:
   a. Circle
   b. OR gate
   c. Rectangle
   d. AND gate
7. Describe TOR.
8. Compare HAZOP analysis with FMEA.
9. Assume that a robot system maintenance person performs three independent safety-related tasks: $x$, $y$, and $z$. Each of these three tasks can be performed either correctly or incorrectly and task $x$ is performed before task $y$, and task $y$ before task $z$.

   Develop a probability-tree and obtain an expression for the probability of not successfully accomplishing the overall mission by the robot system maintenance person.
10. Describe ISA.

## References

1. Countinho, W.E., *Failure Effect Analysis*, Trans. New York Academy of Sciences, Series II, 1963–1964, pp. 564–584.
2. Dhillon, B.S., Singh, C., *Engineering Reliability: New Techniques and Applications*, John Wiley and Sons, New York, 1981.
3. United States Nuclear Regulatory Commission, *Fault Tree Handbook, Report No. NUREG-0492*, Washington, DC, 1981.
4. Hammer, W., Price, D., *Occupational Safety Management and Engineering*, Prentice-Hall, Upper Saddle River, New Jersey, 2001.
5. Dhillon, B.S., *Design Reliability: Fundamentals and Applications*, CRC Press, Boca Raton, Florida, 1999.
6. Dhillon, B.S., *Engineering Safety: Fundamentals, Techniques, and Applications*, World Scientific Publishing, River Edge, New Jersey, 2003.
7. Dhillon, B.S., *Patient Safety: An Engineering Approach*, CRC Press, Boca Raton, Florida, 2012.
8. Dhillon, B.S., *Human Reliability: With Human Factors*, Pergamon Press, New York, 1986.
9. Dhillon, B.S., *Human Reliability, Error, and Human Factors in Engineering Maintenance: With Reference to Aviation and Power Generation*, CRC Press, Boca Raton, Florida, 2009.
10. Omdahl, T.P., Editor, *Reliability, Availability, and Maintainability (RAM) Dictionary*, American Society for Quality Control (ASQC) Press, Milwaukee, Wisconsin, 1988.

11. Bureau of Naval Weapons, U.S. Department of the Navy, *General Specification for Design, Installation, and Test of Aircraft Flight Control Systems*, MIL-F-18372 (Aer), Para. 3.5.2.3., Washington, DC.
12. U.S. Department of Defense, *Procedures for Performing a Failure Mode, Effects, and Criticality Analysis*, MIL-STD-1629, Washington, DC, 1980.
13. Jordan, W.E., Failure modes, effects, and criticality analyses, *Proceedings of the Annual Reliability and Maintainability Symposium*, 1972, pp. 30–37.
14. Palady, P., *Failure Modes and Effects Analysis*, PT Publications, West Palm Beach, Florida, 1995.
15. Shooman, M.L., *Probabilistic Reliability: An Engineering Approach*, McGraw Hill Book Company, New York, 1968.
16. Hallock, R.G., Technique of operations review analysis: Determines causes of accident/incident, *Safety and Health*, Vol. 60, No. 8, 1991, pp. 38–39.
17. Goetsch, D.L., *Occupational Safety and Health*, Prentice-Hall, Englewood Cliffs, New Jersey, 1996.
18. *Guidelines for Hazard Evaluation Procedures*, American Institute of Chemical Engineers, New York, 1985.
19. Dhillon, B.S., *Reliability, Quality, and Safety for Engineers*, CRC Press, Boca Raton, Florida, 2005.
20. Roland, H.E., Moriarty, B., *System Safety Engineering and Management*, John Wiley and Sons, New York, 1983.
21. Gloss, D.S., Wardle, M.G., *Introduction to Safety Engineering*, John Wiley and Sons, New York, 1984.
22. Canadian Standards Association, *Risk Analysis Requirements and Guidelines, Report No. CAN/CSA-Q 634-91*, 1991; available from the Canadian Standards Association, 178 Rexdale Blvd., Rexdale, Ontario, Canada.
23. Hammer, W., *Product Safety Management and Engineering*, Prentice-Hall, Englewood Cliffs, New Jersey, 1980.
24. Swain, A.D., An error-cause removal program for industry, *Human Factors*, Vol. 12, 1973, pp. 207–221.

# 5

# *Robot Reliability*

## 5.1 Introduction

The topic of robot reliability is quite complicated as there are many interlocking variables in evaluating and accomplishing numerous reliability levels. A robot installation is only considered successful if it is safe and reliable. A robot with poor reliability results in problems such as high maintenance-related cost, inconvenience, and unsafe conditions.

Needless to say, the *American National Standard for Industrial Robots and Robot Systems—Safety Requirements* developed by the American National Standards Institute (ANSI) clearly states that the design and construction of robots has to be in such a way that any single, reasonably foreseeable malfunction will not result in any hazardous motion of the robot. As many different types of parts (e.g., electrical, electronic, mechanical, hydraulic, and pneumatic) are used in robots, this makes the task of producing highly reliable robots very challenging. Furthermore, the environments in which the robots have to function may be quite harsh and can vary quite significantly from one application to another even for identical models.

This chapter presents various important aspects of robot reliability.

## 5.2 Classifications of Robot Failures and Their Causes and Corrective Measures

Robot failures can be categorized under the following four classifications [1–3]:

- *Classification I: Random component failures.* These are failures that occur unpredictably during the useful life of components. Some of the reasons for the occurrence of such failures are low safety factors, undetectable defects, unavoidable failures, and unexplainable causes. Some of the methods presented in Chapter 4 and in [4] can be used to reduce the occurrence of such failures.

- *Classification II: Human errors.* These are due to personnel who design, manufacture, test, operate, and maintain robots. Some of the causes for the occurrence of human errors are as follows:
  - Poor training of operating and maintenance personnel
  - Poor equipment design
  - Improper tools
  - Task complexities
  - Poorly written operating and maintenance procedures
  - Inadequate lighting in the work area
  - High temperature in the work area

  Thus, human errors may be divided into categories such as design errors, operating errors, maintenance errors, inspection errors, installation errors, and assembly errors. Some of the methods that can be used to reduce the occurrence of human errors are fault tree analysis, error cause removal program, man–machine systems analysis, and quality control circles. The first method (i.e., fault-tree analysis) is described in Chapter 4 and the remaining three methods are described in [5].
- *Classification III: Systematic hardware faults.* These are failures that occur due to unrevealed mechanisms present in the root system design. Some of the reasons for the occurrence of such faults are peculiar wrist orientations, unusual joint-to-straight-line mode transition, and failure to make the necessary environment-related provisions in the initial design.

  Some of the methods that can be used to reduce the occurrence of robot-related systematic hardware failures are the inclusion of sensors in the system to detect the loss of pneumatic pressure, line voltage, or hydraulic pressure; and the employment of sensors to detect excessiveness of force, speed, servo errors, acceleration, and temperature.

  Several methods useful to reduce systematic hardware failures are described in [4,6].
- *Classification IV: Software failures/errors.* These are associated with software concerned with robots. In robots, software faults/errors/failures can happen in the embedded software or the controlling software and application software. As per one study reported in [7], over 60% of software errors are made during the requirement and design phase of the process as opposed to less than 40% during the coding phase or process.

Redundancy, even though it is expensive, is probably the best solution to protect against software errors or failures. Also, the application of approaches such as failure mode and effect analysis, testing, and fault tree analysis can be helpful to reduce software failures/errors. Furthermore, there are many software reliability models that can be used to evaluate reliability when the software in question is put into operational use [4,6–8].

## 5.3 Robot Effectiveness Dictating Factors and Robot Reliability Survey Results

There are many factors dictating effectiveness of robots. Some of these factors are as follows [3,9]:

- The robot's mean time between failures
- The robot's mean time to repair
- The percentage of time the robot is available for operations
- The percentage of time the robot operates normally
- Quality and availability of the robot repair equipment and facilities
- Rate of the availability of the required spare parts or components
- Quality and availability of personnel required for keeping the robot in operating state
- The relative performance of the robot under extreme conditions

Reference [10] reported the results of a study of robot reliability based on surveys of 37 robots of four different designs used in three different companies A, B, and C; covering 21, 932 robot production hours. These companies (i.e., A, B, and C) reported 47, 306, 155 cases of reliability-related problems of robots, respectively, of which the corresponding 27, 35, and 1 cases did not contribute to downtime. More clearly, robot downtime as a proportion of production time for these companies (i.e., A, B, and C) was 1.8%, 13.6%, and 5.1%, respectively.

Approximate mean time to robot failure (*MTTRF*) and mean time to robot-related problems (*MTTRP*) in hours for companies A, B, and C are presented in Table 5.1.

It is to be noted that among these three companies, there is a wide variation of *MTTRF* and *MTTRP*. More clearly, the lowest and highest *MTTRF* and *MTTRP* levels are 40, 2596, 15 and 221 h, respectively.

**TABLE 5.1**

Approximate Figures for Mean Time to Robot Failure (*MTTRF*) and Mean Time to Robot-Related Problems (*MTTRP*)

| No. | Company | *MTTRF* (h) | *MTTRP* (h) |
|---|---|---|---|
| 1 | A | 2596 | 221 |
| 2 | B | 284 | 30 |
| 3 | C | 40 | 15 |

## 5.4 Robot-Related Reliability Measures

There are various types of robot-related reliability measures. Four of these are presented below [3,4,11].

### 5.4.1 Mean Time to Robot-Related Problems

This is the average productive robot time before the occurrence of a robot-related problem and is expressed by

$$MTTRP = \frac{RPH - DDTRP}{NRP} \tag{5.1}$$

where

*MTTRP* is the mean time to robot-related problems.
*RPH* is the robot production hours.
*NRP* is the number of robot-related problems.
*DDTRP* is the downtime due to robot-related problems expressed in hours.

**EXAMPLE 5.1**

Assume that the annual robot production hours and downtime due to robot-related problems, at an industrial facility, are 6000 and 500 h, respectively. During the 1-year period, there were 25 robot-related problems. Calculate the average time to robot-related problems.

By substituting the given data values into Equation 5.1, we get

$$\begin{aligned} \text{MTTRP} &= \frac{6000 - 500}{25} \\ &= 220 \text{ h} \end{aligned}$$

Thus, the average time to robot-related problems is 220 h.

### 5.4.2 Mean Time to Robot-Related Failure

Mean time to robot-related failure can be obtained by using either of the following three expressions:

$$MTTF_r = \frac{RPH - DDTRF}{NRF} \tag{5.2}$$

$$MTTF_r = \int_0^\infty R_r(t)dt \tag{5.3}$$

$$MTTF_r = \lim_{s \to 0} R_r(s) \tag{5.4}$$

where

$MTTF_r$ is the robot's mean time to failure.
$DDTRF$ is the downtime due to robot-related failure expressed in hours.
$NRF$ is the number of robot failures.
$R_r(t)$ is the robot reliability at time $t$.
$s$ is the Laplace transform variable.
$R_r(s)$ is the Laplace transform of the robot reliability function, $R_r(t)$.

**EXAMPLE 5.2**

Assume that the annual production hours of a robot and its annual downtime due to failures are 4000 and 200 h, respectively. During that period, the robot failed five times. Calculate the mean time to robot-related failure.

By inserting the specified data values into Equation 5.2, we obtain

$$\begin{aligned} \text{MTTF}_r &= \frac{4000 - 200}{5} \\ &= 760 \text{ h} \end{aligned}$$

Thus, the mean time to robot-related failure is 760 h.

**EXAMPLE 5.3**

Assume that the constant failure rate, $\lambda_r$, of a robot is 0.0002 failures per hour and its reliability is expressed by

$$\begin{aligned} R_r(t) &= e^{-\lambda_r t} \\ &= e^{-(0.0002)t} \end{aligned} \tag{5.5}$$

where

$R_r(t)$ is the robot reliability at time $t$.

Calculate the mean time to robot-related failure with the aid of Equations 5.3 and 5.4 and comment on the final result.

By inserting Equation 5.5 into Equation 5.3, we get

$$\begin{aligned} MTTF_r &= \int_0^\infty e^{-(0.0002)t}\,dt \\ &= \frac{1}{0.0002} \\ &= 5000 \text{ h} \end{aligned}$$

By taking the Laplace transform of Equation 5.5, we obtain

$$R_r(s) = \frac{1}{(s + 0.0002)} \tag{5.6}$$

By substituting Equation 5.6 into Equation 5.4, we get

$$\begin{aligned} MTTF_r &= \lim_{s \to 0} \frac{1}{(s + 0.0002)} \\ &= \frac{1}{0.0002} \\ &= 5000 \text{ h} \end{aligned}$$

In both cases, the final result (i.e., $MTTF_r = 5000$ h) is the same. It proves that Equations 5.3 and 5.4 yield exactly the same result.

### 5.4.3 Robot Reliability

Robot reliability may simply be described as the probability that a robot will perform its specified function satisfactorily for the stated time interval when used as per designed conditions. The general formula to obtain time-dependent robot reliability is expressed by [3,4]

$$R_r(t) = \exp\left[-\int_0^t \lambda_r(t)\,dt\right] \tag{5.7}$$

where

$R_r(t)$ is the robot reliability at time $t$.

$\lambda_r(t)$ is the robot's time-dependent failure rate (hazard rate).

Equation 5.7 can be used for obtaining the reliability function of a robot for any failure times probability distribution (e.g., Weibull, gamma, or exponential).

**EXAMPLE 5.4**

Assume that the time-dependent failure rate of a robot is expressed by

$$\lambda_r(t) = \frac{bt^{b-1}}{\gamma\gamma^{b-1}} \tag{5.8}$$

where

$b$ is the shape parameter.
$t$ is time.
$\gamma$ is the scale parameter.
$\lambda_r(t)$ is the hazard rate or time-dependent failure rate of the robot when its times to failure follows the Weibull distribution.

Obtain an expression for the robot reliability.

By substituting Equation 5.8 into Equation 5.7, we get

$$R_r(t) = \exp\left[-\int_0^t \frac{bt^{b-1}}{\gamma\gamma^{b-1}}dt\right]$$

$$= e^{-(t/\gamma)^b} \tag{5.9}$$

Thus, Equation 5.9 is the expression for the robot reliability.

**EXAMPLE 5.5**

Assume that the constant failure rate of a robot is 0.0002 failures per hour. Calculate its reliability for a 15-h mission.

By substituting the given constant failure rate value of the robot into Equation 5.7, we obtain

$$R_r(t) = \exp\left[-\int_0^t (0.0002)dt\right]$$

$$= e^{-(0.0002)t} \tag{5.10}$$

Inserting the given mission time value of the robot into Equation 5.10 yields

$$R_r(15) = e^{-(0.0002)(15)}$$

$$= 0.9970$$

Thus, the robot reliability for the specified mission period is 0.9970.

### 5.4.4 Robot's Hazard Rate

The robot's hazard rate or time-dependent failure rate is defined by [3,4]

$$\lambda_r(t) = -\frac{1}{R_r(t)} \cdot \frac{dR_r(t)}{dt} \tag{5.11}$$

where

$R_r(t)$ is the robot reliability at time $t$.
$\lambda_r(t)$ is the robot's hazard rate (time-dependent failure rate).

Equation 5.11 can be used for obtaining the robot's hazard rate when the robot's times to failure follows any time-continuous probability distribution (e.g., Weibull, gamma, and Rayleigh, exponential).

**EXAMPLE 5.6**

Assume that the reliability of a robot is expressed by

$$R_r(t) = e^{-\lambda_r t} \tag{5.12}$$

where

$R_r(t)$ is the robot reliability at time $t$.
$\lambda_r$ is the robot's constant failure rate.

Obtain an expression for the robot's hazard rate.

By substituting Equation 5.12 into Equation 5.11, we get

$$\begin{aligned} \lambda_r(t) &= -\frac{1}{e^{-\lambda_r t}} \cdot \frac{de^{-\lambda_r t}}{dt} \\ &= -\frac{1}{e^{-\lambda_r t}}[-\lambda_r e^{-\lambda_r t}] \\ &= \lambda_r \end{aligned} \tag{5.13}$$

Equation 5.13 is the expression for the robot's hazard rate, which is constant (i.e., it does not depend on time).

## 5.5 Robot Reliability Analysis Methods and Models for Performing Robot Reliability Studies

In the field of reliability engineering, there are many methods used to perform various types of reliability analyses. Some of these methods can be used quite effectively to perform robot reliability analysis. Four of these methods are as follows:

- *Failure modes and effect Analysis (FMEA).* This method was developed by the U.S. Department of Defense in the early 1950s and is considered an effective tool to perform analysis of each failure mode in the equipment/system for determining the effects of such failure modes on the entire equipment/system [12].

  The method is composed of six steps: (i) define equipment/system and its associated requirements, (ii) develop necessary ground rules, (iii) describe the equipment/system and all its associated functional blocks, (iv) identify all possible failure modes and their effects, (v) develop a list of critical items, and (vi) document the analysis [12–14]. Additional information on FMEA is available in Chapter 4 and in [4,15].
- *Fault tree analysis (FTA).* This method was developed at the Bell Telephone Laboratories in the early 1960s and is widely used to evaluate reliability of engineering systems during their design and development phase. A fault tree may simply be described as a logical representation of the relationship of primary/basic fault events that lead to the occurrence of a specified undesired event called the "top event."

  Additional information on fault tree analysis is available in Chapter 4 and in [4,16].
- *Markov method.* This method can be used in more cases than any other reliability evaluation method. It is concerned with modeling systems with constant failure and repair rates.

  Additional information on this method is available in Chapter 4 and in [4,17].
- *Parts count method.* This method is generally used during a bid proposal and early design phases to estimate the equipment/system failure rate. The method requires information on items such as quality levels of parts, quantities and types of generic parts, and the use environment of equipment/system/products.

  Additional information on the method is available in [4,18].

Over the years, many mathematical models have been developed to perform reliability-related studies of robots. Although the effectiveness of these models can vary significantly from one application area to another, some are being used quite successfully for performing reliability-related studies of robots in the industrial sector.

Many of these mathematical models were developed using the Markov method. Additional information about these models is available in Chapter 11 and in [3,11,19–21].

## 5.6 Reliability Analysis of Hydraulic and Electric Robots

As both hydraulic and electric robots are used in the industrial sector, this section presents reliability analyses of two typical hydraulic and electric robots by using the block diagram method [1–3]. Usually, for the purpose of design evaluation, it is assumed for both hydraulic and electric robots that all robot parts act in series (i.e., if any one part fails, the robot fails).

### 5.6.1 Reliability Analysis of the Hydraulic Robot

A hydraulic robot considered here contains five joints and, in turn, each joint is driven and controlled by a hydraulic servomechanism. The robot is subject to the following assumptions/factors [1,3]:

- Under high flow demand, an accumulator assists the pump to supply additional hydraulic fluid.
- The position transducer provides the joint angle codes and, in turn, the scanning of each code is carried out by a multiplexer.
- A conventional motor and pump assembly produces pressure.
- An operator makes use of a teach pendant for controlling the arm-motion in teach mode.
- Hydraulic fluid is pumped from the reservoir.
- The motion of each hydraulic actuator is controlled by a servo valve. This motion is transmitted directly or indirectly (i.e., through rods, chains, gears, etc.) to the specific limb of the robot and, in turn, each limb is coupled to a position transducer.
- An unloading valve is used for keeping pressure under the maximum limit.

The block diagram shown in Figure 5.1 represents the hydraulic robot under consideration with respect to reliability. Figure 5.1 shows that the hydraulic robot is composed of four subsystems (i.e., gripper subsystem, hydraulic pressure supply subsystem, drive subsystem, and electronic and

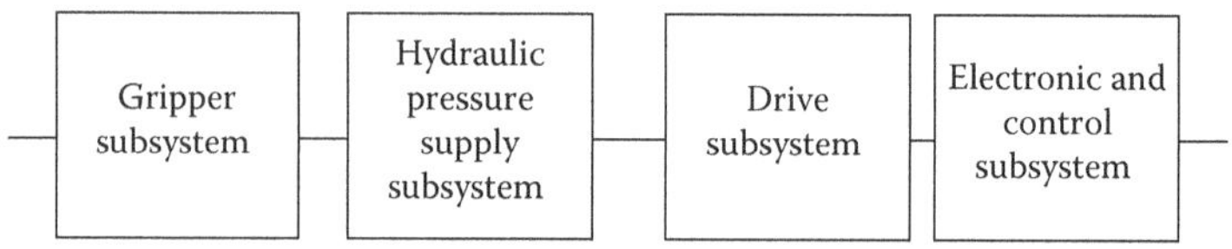


**FIGURE 5.1**
Block diagram of the hydraulic robot under consideration.

control subsystem) in series. In turn, as shown in the Figure 5.2 gripper subsystem (i.e., block diagram (a)) is composed of two parts (i.e., control signal and pneumatic system) in series, hydraulic pressure supply subsystem (i.e., block diagram (b)) is also composed of two parts (i.e., piping and hydraulic equipment) in series, and drive subsystem (i.e., block diagram (c)) is composed of five parts (i.e., joints 1, 2, 3, 4, and 5) in series.

With the aid of Figure 5.1, we have the following expression for the probability of the nonoccurrence of the hydraulic robot event (i.e., undesirable hydraulic robot arm movement causing damage to the robot/other equipment and possible harm to humans):

$$R_{hr} = R_1 R_2 R_3 R_4 \tag{5.14}$$

where

$R_{hr}$ is the hydraulic robot reliability or the probability of the nonoccurrence of the hydraulic robot event (i.e., undesirable robotic arm movement causing damage to the robot/other equipment and possible harm to humans).

$R_1$ is the reliability of the independent gripper subsystem.

$R_2$ is the reliability of the independent hydraulic pressure supply subsystem.

$R_3$ is the reliability of the independent drive subsystem.

$R_4$ is the reliability of the independent electronic and control subsystem.

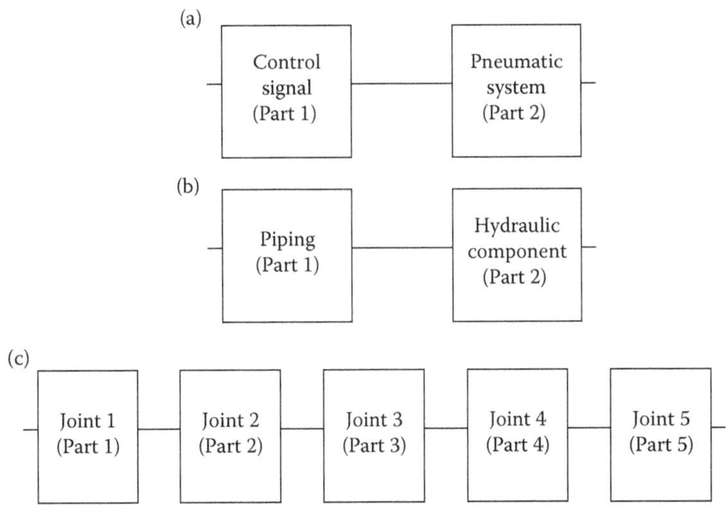


**FIGURE 5.2**
Block diagram representing three subsystems shown in Figure 5.1: (a) gripper subsystem, (b) hydraulic pressure supply subsystem, and (c) drive subsystem.

For independent parts, the reliabilities $R_1$, $R_2$, and $R_3$ of gripper subsystem, hydraulic pressure supply subsystem, and drive subsystem using Figure 5.2a–c, respectively, are

$$R_1 = R_{cs}R_{ps} \tag{5.15}$$

$$R_2 = R_pR_{hc} \tag{5.16}$$

and

$$R_3 = \prod_{i=1}^{5} R_{ji} \tag{5.17}$$

where
$R_1$ is the reliability of the gripper subsystem.
$R_2$ is the reliability of the hydraulic pressure supply subsystem.
$R_3$ is the reliability of the drive subsystem.
$R_{cs}$ is the reliability of the control signal.
$R_{ps}$ is the reliability of the pneumatic system.
$R_p$ is the reliability of the piping.
$R_{hc}$ is the reliability of the hydraulic component.
$R_{ji}$ is the reliability of joint $i$, for $i = 1,2,3,4,5$.

For constant failure rates of independent subsystems shown in Figure 5.1, and, in turn, of their corresponding independent parts shown in Figure 5.2; from Equations 5.14 through 5.17, we obtain

$$\begin{aligned} R_{hr}(t) &= e^{-\lambda_1 t}e^{-\lambda_2 t}e^{-\lambda_3 t}e^{-\lambda_4 t} \\ &= e^{-\lambda_{cs} t}e^{-\lambda_{ps} t}e^{-\lambda_p t}e^{-\lambda_{hc} t}e^{-\sum_{i=1}^{5}\lambda_i t}e^{-\lambda_4 t} \\ &= e^{-(\lambda_{cs}+\lambda_{ps}+\lambda_p+\lambda_{hc}+\sum_{i=1}^{5}\lambda_{ji}+\lambda_4)t} \end{aligned} \tag{5.18}$$

where
$R_{hr}(t)$ is the hydraulic robot reliability or the probability of the nonoccurrence of the hydraulic robot event (i.e., undesirable robotic arm movement, causing damage to the robot/other equipment and possible harm to humans) at time $t$.
$\lambda_1$ is the constant failure rate of the gripper subsystem.
$\lambda_2$ is the constant failure rate of the hydraulic pressure supply subsystem.
$\lambda_3$ is the constant failure rate of the drive subsystem.
$\lambda_4$ is the constant failure rate of the electronic and control subsystem.
$\lambda_{cs}$ is the constant failure rate of the control signal (part 1).

$\lambda_{ps}$ is the constant failure rate of the pneumatic system (part 2).
$\lambda_p$ is the constant failure rate of the piping (part 1).
$\lambda_{hc}$ is the constant failure rate of the hydraulic component (part 2).
$\lambda_{ji}$ is the constant failure rate of the joint $i$, for $i = 1,2,3,4,5$.

By integrating Equation 5.18 over the time interval $[0,\infty]$, we get

$$MTTOHRE = \int_0^\infty e^{-(\lambda_{cs}+\lambda_{ps}+\lambda_p+\lambda_{hc}+\sum_{i=1}^5 \lambda_{ji}+\lambda_4)t} dt$$

$$= \frac{1}{\lambda_{cs} + \lambda_{ps} + \lambda_p + \lambda_{hc} + \sum_{i=1}^5 \lambda_{ji} + \lambda_4} \tag{5.19}$$

where

*MTTOHRE* is the mean time to the occurrence of the hydraulic robot event: undesirable arm movement causing damage to the robot/other equipment and possible harm to humans.

**EXAMPLE 5.7**

Assume that the following reliability data values are given for the above type of hydraulic robot:

$$R_{j1} = R_{j2} = R_{j3} = R_{j4} = R_{j5} = 0.92, \ R_4 = 0.95,$$

$$R_{cs} = 0.86, \ R_{ps} = 0.96, \ R_p = 0.94, \quad \text{and} \quad R_{hc} = 0.9$$

Calculate the non occurrence probability of the event: undesirable hydraulic robot arm movement causing damage to the robot/equipment and possible harm to humans.

By substituting the given data values into Equations 5.15 through 5.17, we get

$$R_1 = (0.86)(0.96) = 0.8256$$

$$R_2 = (0.94)(0.9) = 0.846$$

and

$$R_3 = (0.92)(0.92)(0.092)(0.92)(0.92) = 0.6591$$

By substituting the above-calculated values and the given data value into Equation 5.14, we get

$$R_{hr} = (0.8256)(0.846)(0.6591)(0.95)$$

$$= 0.4373$$

Thus, the non occurrence probability of the event: undesirable hydraulic robot arm movement causing damage to the robot/equipment and possible harm to humans, is 0.4373.

**EXAMPLE 5.8**

Assume that the specified constant failure rate values of the above type of hydraulic robot are $\lambda_{cs} = 0.0007$ failures per hour, $\lambda_{ps} = 0.0006$ failures per hour, $\lambda_p = 0.0005$ failures per hour, $\lambda_{hc} = 0.0004$ failures per hour, $\lambda_{j1} = \lambda_{j2} = \lambda_{j3} = \lambda_{j4} = \lambda_{j5} = 0.0003$ failures per hour, and $\lambda_4 = 0.0002$ failures per hour, respectively.

Using Equation 5.19, calculate the mean time to the occurrence of the hydraulic robot event: undesirable arm movement causing damage to the robot/other equipment and possible harm to humans.

By substituting the specified data values into Equation 5.19, we get

$$\begin{aligned} MTTOHRE &= \frac{1}{0.0007 + 0.0006 + 0.0005 + 0.0004 + 5(0.0003) + 0.0002} \\ &= 256.41 \text{ h} \end{aligned}$$

Thus, the mean time to the occurrence of the hydraulic robot event: undesirable robot arm movement causing damage to the robot/other equipment and possible harm to humans is 256.41 h.

### 5.6.2 Reliability Analysis of the Electric Robot

An electric robot considered here is one that performs a "normal" industrial task, while its programming and maintenance are carried out by humans. The robot is subject to the following assumptions/factors [2,3]:

- The motor shaft rotation is transmitted to the appropriate limb of the robot through a transmission unit.
- A microprocessor control card controls each and every joint.
- A supervising computer/controller directs all joints.
- A direct current (DC) motor actuates each joint.
- An interface bus allows interaction between the supervisory controller and the joint control processors.
- The transducer sends all appropriate signals to the joint controller.
- Each and every joint is coupled with a feedback transducer (encoder).

The block diagram shown in Figure 5.3 represents the electric robot under consideration with respect to reliability. Figure 5.3 shows that the electric robot is composed of two hypothetical subsystems (i.e., no movement due to external factors (subsystem 1) and no failure within the robot causing its

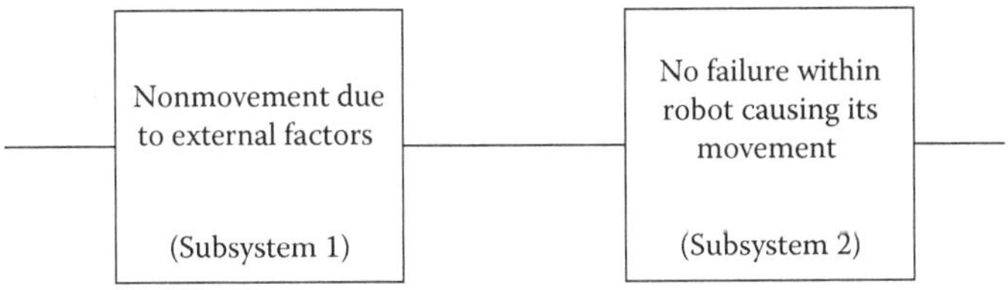


**FIGURE 5.3**
Block diagram for estimating the nonoccurrence probability (reliability) of the undesirable movement of the electric robot.

movement (subsystem 2) in series. In turn, as shown in the Figure 5.4, subsystem 1 (i.e., block diagram (a)) is composed of two hypothetical elements (i.e., element *A* and element *B*) in series and subsystem 2 (i.e., block diagram (c)) is composed of five parts (i.e., interface, end effector, supervisory computer/controller, joint control, and drive transmission) in series.

Furthermore, as shown in Figure 5.4, element *A* (i.e., block diagram (b)) is composed of two hypothetical subelements (i.e., subelement *X* and subelement *Y*) in series.

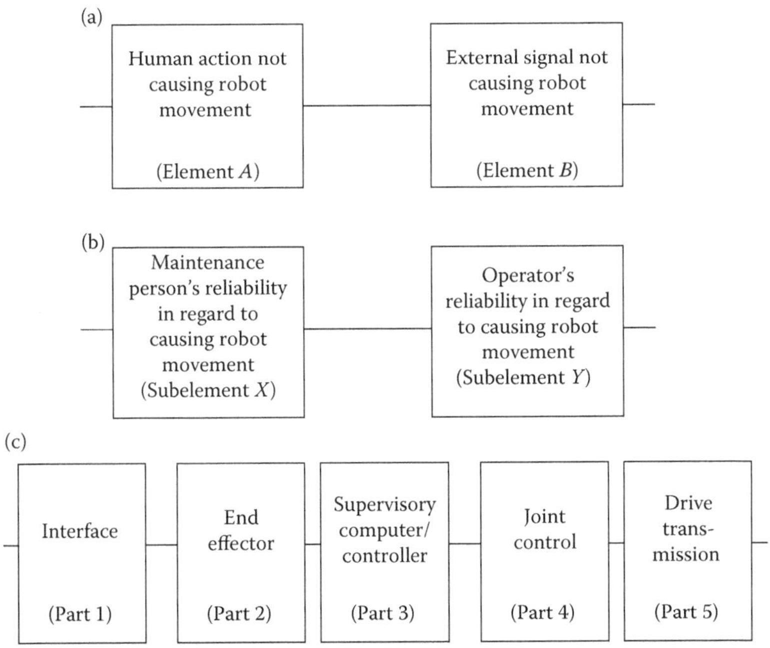


**FIGURE 5.4**
Block diagram representing two subsystems shown in Figure 5.3: (a) subsystem 1, (b) element *A* of subsystem 1, and (c) subsystem 2.

With the aid of Figure 5.3, we have the following expression for the probability of nonoccurrence of the undesirable electric robot movement (reliability):

$$R_{er} = R_1 R_2 \tag{5.20}$$

where

$R_{er}$ is the probability of nonoccurrence (reliability) of the undesirable electric robot movement.

$R_1$ is the reliability of the independent subsystem 1.

$R_2$ is the reliability of the independent subsystem 2.

For independent elements, the reliability of subsystem 1 in Figure 5.4a is

$$R_1 = R_A R_B \tag{5.21}$$

where

$R_A$ is the reliability of element $A$.

$R_B$ is the reliability of element $B$.

For independent and hypothetical subelements, the reliability of element $A$ in Figure 5.4b is

$$R_A = R_X R_Y \tag{5.22}$$

where

$R_X$ is the maintenance person's reliability with regard to causing the robot's movement (subelement $X$).

$R_Y$ is the operator's reliability with regard to causing the robot's movement (subelement $Y$).

Similarly, for independent parts, the reliability of subsystem 2 in Figure 5.4c is

$$R_2 = R_i R_e R_s R_j R_d \tag{5.23}$$

where

$R_i$ is the reliability of the interface.

$R_e$ is the reliability of the end-effector.

$R_s$ is the reliability of the supervisory computer/controller.

$R_j$ is the reliability of the joint control.

$R_d$ is the reliability of the drive transmission.

**EXAMPLE 5.9**

Assume that the following reliability data values are given for an above type of electric robot:

$$R_B = 0.96,\ R_X = 0.91,\ R_Y = 0.92,\ R_i = 0.90,$$
$$R_e = 0.93,\ R_s = 0.94,\ R_j = 0.95,\quad \text{and}\quad R_d = 0.97$$

Calculate the probability of nonoccurrence (reliability) of the undesirable electric robot movement.

By substituting the specified data values into Equations 5.22 and 5.23, we get

$$R_A = (0.91)(0.92) = 0.8372$$

and

$$R_2 = (0.90)(0.93)(0.94)(0.95)(0.97) = 0.7250$$

By inserting the above-calculated value for $R_A$ and the given data value for $R_B$ into Equation 5.21, we get

$$R_1 = (0.8372)(0.96) = 0.8037$$

By substituting the above-calculated values into Equation 5.20, we obtain

$$R_{er} = (0.8037)(0.7250) = 0.5827$$

Thus, the probability of nonoccurrence (reliability) of the undesirable electric robot movement is 0.5827.

## 5.7 Problems

1. Discuss classifications of robot failures and their causes.
2. What are the dictating factors for robot effectiveness?
3. Write down the formula to estimate mean time to robot-related problems.
4. Assume that the annual production hours of a robot and its annual downtime due to failures are 5000 and 150 h, respectively. During that period the robot failed four times. Calculate the mean time to robot failure.

5. Assume that the constant failure rate, $\lambda_r$, of a robot is 0.0004 failures per hour and its reliability is expressed by

$$R_r(t) = e^{-\lambda_r t} = e^{-(0.0004)t} \tag{5.24}$$

where

$R_r(t)$ is the robot reliability at time $t$.

Calculate the mean time to robot failure with the aid of Equations 5.3 and 5.4 and comment on the end result.
6. Prove Equation 5.9 by using Equations 5.7 and 5.8.
7. Assume that the reliability of a robot is expressed by Equation 5.9. Obtain an expression for the robot's hazard rate.
8. Discuss at least four methods that can be used to analyze robot reliability.
9. Compare a hydraulic robot with an electric robot with respect to reliability.
10. Discuss with results of the robot reliability surveys concerning 37 robots of four different designs used in three different companies, covering almost 22,000 robot production hours.

## References


1. Khodanbandehloo, K., Duggan, F., Husband, T.F., Reliability assessment of industrial robots, *Proceedings of the 14th International Symposium on Industrial Robots*, 1984, pp. 209–220.
2. Khodabandehloo, K., Duggan, F., Husband, T.F., Reliability of industrial robots: A safety viewpoint, *Proceedings of the 7th British Robot Association Annual Conference*, 1984, pp. 233–242.
3. Dhillon, B.S., *Robot Reliability and Safety*, Springer-Verlag, New York, 1991.
4. Dhillon, B.S., *Design Reliability: Fundamentals and Applications*, CRC Press, Boca Raton, Florida, 1999.
5. Dhillon, B.S., *Human Reliability: With Human Factors*, Pergamon Press, New York, 1986.
6. Dhillon, B.S., *Reliability Engineering in Systems Design and Operation*, Van Nostrand Reinhold Company, New York, 1983.
7. Dhillon, B.S., *Reliability in Computer Systems Design*, Ablex Publishing, Norwood, New Jersey, 1987.
8. Herrmann, D.S., *Software Safety and Reliability*, IEEE Computer Society Press, Los Alamitos, California, 1999.
9. Young, J.F., *Robotics*, Butterworth, London, 1973.

10. Jones, R., Dawson, S., People and robots: Their safety and reliability, *Proceedings of the 7th British Robot Association Annual Conference*, 1984, pp. 243–258.
11. Dhillon, B.S., *Applied Reliability and Quality: Fundamentals, Methods, and Procedures*, Springer, London, 2007.
12. Omdahl, T.P., Editor, *Reliability, Availability, and Maintainability (RAM) Dictionary*, American Society for Quality Control (ASQC) Press, Milwaukee, Wisconsin, 1988.
13. Coutinho, J.S., Failure effect analysis, *Transactions of the New York Academy of Sciences*, Vol. 26, Series II, 1963–1964, pp. 564–584.
14. Bureau of Naval Weapons, Department of the Navy, *General Specification for Design, Installation, and Test of Aircraft Flight Control Systems, MIL-F-18372 (Aer)*, Washington, DC, Para. 3.5.2.3.
15. Palady, P., *Failure Modes and Effects Analysis*, PT Publications, West Palm Beach, Florida, 1995.
16. U.S. Nuclear Regulatory Commission, *Fault-Tree Handbook, Report No. NUREG-0492*, Washington, DC, 1981.
17. Shooman, M.L., *Probabilistic Reliability: An Engineering Approach*, McGraw-Hill Book Company, New York, 1968.
18. US Department of Defense, *Reliability Prediction of Electronic Equipment, MIL-HDBK-217*, Washington, DC.
19. Dhillon, B.S., Fashandi, A.R.M., Robotic systems probabilistic analysis, *Microelectronics and Reliability*, Vol. 37, 1997, pp. 211–224.
20. Dhillon, B.S., Fashnadi, A.R.M., Stochastic analysis of a robot machine with duplicate safety units, *Journal of Quality in Maintenance Engineering*, Vol. 5, No. 2, 1999, pp. 114–127.
21. Dhillon, B.S., Li, Z., Stochastic analysis of a system with redundant robots, one built-in safety unit, and common cause failures, *Journal of Intelligent and Robotics Systems*, Vol. 45, 2006, pp. 137–155.

# 6

# *Robot Safety*

## 6.1 Introduction

Today, robots are being used in many diverse areas and applications, and their safety-related problems have increased quite significantly. Each area and application may call for specific precautions for programmers, operators, robot systems, maintenance workers, and so on. In order to meet needs such as these, various organizations have, over the years, developed documents that are specifically concerned with robot safety.

Two examples of such documents are the *American National Standard for Industrial Robots and Robot Systems—Safety Requirements* [1] and the Japanese Industrial Safety and Health Association document titled, "An Interpretation of the Technical Guidance on Safety Standards in the Use, Etc., of Industrial Robots" [2].

Generally, factors such as replacing humans in performing hazardous and difficult tasks and increasing productivity play an important role in the use of robots in the industrial sector. Improper consideration to safety in robotics planning may create hazardous conditions other than those that the robot may be replacing. It simply means that for successful applications of robots, safety must be considered with utmost care during the planning phase.

This chapter presents various important aspects of robot safety.

## 6.2 Robot Safety: Problems and Hazards

Safety professionals concerned with robots face many unique robot-related safety problems. Some of these problems are as follows [3,4]:

- Robots generate potentially hazardous conditions because they manipulate items of varying weights and sizes.
- The presence of a robot receives great attention from humans in the surrounding area, who are often quite ignorant of the possible associated hazards.

- Robot-related maintenance procedures may lead to hazardous situations.
- A robot may lead to a high risk of fire or an explosion if it is placed in an unfriendly environment.
- In the event of the occurrence of a control, hydraulic, or a mechanical failure, robots may move out of their programmed area zones and strike something or they may throw some small item.
- Normally, robots function quite closely with other machines and humans in the same environment. In particular, humans are subject to collisions with the moving parts of robots, tripping over loose power/control cables (if any), and of being pinned down.
- Various safety-associated electrical design problems can occur in robots. Some examples of these problems are poorly designed power sources, fire hazards, and potential for electric shock.
- Robot mechanical design-related problems may result in hazards such as grabbing, pinning, and pinching.
- Management attitudes very much lead to miscomprehension of robot safety-related concepts.

Three basic types of robot hazards are shown in Figure 6.1 [3–5]—trapping, impact, and the hazards that develop from the application.

The trapping hazard is usually the result of robot movements in regard to fixed objects such as pests and machines in the same area. The movements of the auxiliary equipment could be the other possible causes. Two examples of such equipment are pallets and carriages.

The impact hazard involves being struck by the moving part of a robot or by parts or items being carried by the robot. This hazard also includes being struck by flying objects that are ejected or dropped by the robot.

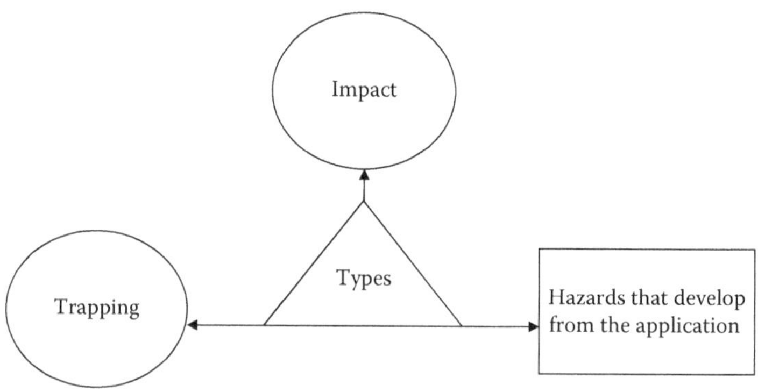


**FIGURE 6.1**
Basic types of robot-related hazards.

Some of the hazards that develop from the application are burns, electric shocks, exposure to toxic substances, and arc flash. The most prevalent causes of these hazards are as follows [6]:

- Control errors
- Mechanical-related problems
- Human error
- Unauthorized access

Control errors may simply be described as intrinsic faults within the robot's control system. Some examples of control errors are electrical interference, faults in the hydraulic, pneumatic, or electrical subcontrol systems, and software problems. The mechanical-related problems arise from the application or the robot system itself. Generally, mechanical hazards occur during the transfer of items with sharp edges, and from malfunctions due to poor maintenance or overloading.

Human error occurs when humans enter the robot's protected work zone and carry out their required tasks close to it. Usually, complacency or lack of care during human interaction leads to human error. Finally, unauthorized access is probably the most prevalent source of hazard; it can easily be controlled by developing proper standards for access, and then following these effectively.

## 6.3 Roles of Robot Manufacturers and Users in Robot Safety

As both robot manufacturers and users profit from robot-related safety, each has a very important role to play in its effective promotion. The basic responsibility of robot manufacturers is to design and manufacture safe robots. The design-related safety measures of robots can be grouped under the following five categories [6–8]:

- *Mechanical subsystem.* Safety in mechanical subsystems of robots can be built in the design process, by taking into consideration the sources of hazards and appropriate measures to eliminate them. Kinetic energy storage capacity, poor reliability, incompatibility of materials with the work environment, and pinch-points are the main sources of hazard in the mechanical subsystem. Furthermore, during the design process, proper attention must be paid to specific sources of danger such as uncovered ball screws, exposed motors, protruding linkages, and flopping cables and hoses.

The robot kinetic energy level can be minimized by optimizing the design for minimum arm inertia. The mechanisms such as dynamic braking, dumping within the drive system, mechanical stops and shock absorbers, and conversion to elastic energy in spring-type parts can be used to attain control of robotic kinetic energy.

- *Electrical/electronic subsystem.* There are various safety-related features that can be designed into the robot's electrical/electronic subsystem, including emergency stops that can be activated by an operator in the event of a hazardous situation, intrinsic safety circuits, input/output signal conditioning, and equipping the teach pendant with a dead-man switch.
- *Operational procedures.* These procedures are an integral part of robot design and they must be provided effectively for the safe operation of the robot. They include items such as operating manuals, sets of instructions, and appropriate precautions.
- *Control subsystem's algorithm.* Robot-related safety can be improved appreciably through the effective control subsystem algorithm design. Therefore, proper attention has to be paid to items such as the regulation and control of the maximum operating speed of the drives and the definition of the tolerable feedback system's following error.
- *Operational and control subsystems' software.* The most safety-related awareness related to the robot can be introduced in the design of the software for the operational and control subsystem. It is to be noted that the capabilities that can be designed into robots are only useful if the logic system is active as well as the robots themselves are calibrated properly. Some examples of the capabilities are
  - Imposition of a maximum robot-operating speed during teaching
  - A response system to switch-off signals from outside interfaces
  - A response system to communication data-flow abnormalities
  - An emergency response system to an unusual velocity on every server-controlled drive
  - A response to interrupts generated by limits associated with hardware travel

Past experiences indicate that the users of robots play an equally important role with regard to robot-related safety. The robot users' safety-related responsibilities may be grouped in the following three categories [7]:

- *The engineering/maintenance department.* The responsibilities of the engineering/maintenance department include installation according to the instructions of the manufacturer, locating all control stations, with the exception of the Pendant Control (whenever required) outside the restricted operating zone; a proper maintenance program, and effective training of maintenance workers.

- *The management.* With regard to safety, the basic two responsibilities of the robot user's management team are in appreciating the safety implications of robotization and supporting the safety department/unit to promote the special features of robot safety.
- *The safety department.* The safety department has many responsibilities, including installing appropriate warning signs or other measures in the robot work envelope, installing barriers with interlocking gates at the work boundary, providing appropriate safety-related training to all concerned personnel, collaborating with other groups concerned with the purchase and the use of robots, and keeping abreast with the latest developments associated with robot-related safety.

## 6.4 Safety Considerations in Robot Design, Installation, Programming, and Operation and Maintenance Phases*

In order to minimize the problems of robot-related safety, it is absolutely essential to carefully consider the safety factor during the robot design, installation, programming, and operation and maintenance phases. Some useful safety-related guidelines concerning each of these four phases are presented below, separately [2,8–11].

### 6.4.1 Robot Design Phase

The robot design safety features may be categorized under the following three classifications:

- *Mechanical.* The mechanical safety features include eliminating sharp corners, designing teach pendant ergonomically, ensuring the existence of mechanisms for releasing the stopped energy, having drive mechanism covers, putting guards on items such as belts, gears, and pulleys, having dynamic brakes for software crash or power failure, and having several emergency stop buttons.

* It is to be noted that the material presented in this section may overlap to a certain degree with the preceding section, but is included here for completeness.

- *Electrical.* The electrical safety features include ensuring the internal safety of the robot so that it will not ignite in a combustible environment, having built-in hose and cable routes using adequate insulation, sectioning and providing panel covers, designing wire circuitry capable of stopping the robot's movement and locking its brakes, minimizing the effects of electromagnetic and radiofrequency interferences, having a fuse "blow" long before human crushing pressure is experienced, and eliminating the risk of an electric shock.
- *Software.* The software safety features include using a procedure of checks to determine why a failure occurred, having built-in commands, periodically examining the built-in self-checking software for safety, having a standby power source for the robot's functioning with programs in random access memory, prohibiting a restart by merely resetting a single switch, providing a robot motion simulator, and having a restart approach after experiencing an emergency stop.

### 6.4.2 Robot Installation Phase

There are many safety features that related to the robot installation phase. Some of these features are as follows:

- Installing electrical cables according to electrical codes
- Installing appropriate interlocks, sensing devices, and so on
- Providing an appropriate level of illumination to humans concerned with the robot
- Placing robot controls outside the hazard area
- Identifying the danger zones with the aid of codes, signs, line markings, and so on
- Ensuring the accessibility and visibility of emergency stops
- Placing an appropriate shield between the robot and humans
- Providing protection to control circuitry by filtering spikes and surges
- Using vibration-reducing pads when required
- Adding pads, cushions, and so on to possible collision points with humans
- Controlling environmental factors as necessary
- Labeling the sources of stored energy
- Distancing circuit boards from electromagnetic fields
- Installing the required interlocks to interrupt robot motion

### 6.4.3 Robot Programming Phase

Safety during the robot programming phase is as important as during its design and installation phases. As per [10], a study of 131 cases reported that programmers/setters were at the highest risk, accounting for 57% of the accidents; the comparative figures for fault clearance personnel, maintenance personnel (repair/servicing), and operators were 26%, 4%, and 13%, respectively.

There are various factors that characterize a robot programmer's work conditions, including working in a bending position (i.e., normally in the robot's movement zone), frequently changing position because the torch's tip is concealed by the clamping device, and being subject to stress by improper lighting. Some of the safety measures that can be taken into consideration with respect to robot-related safety are as follows [10]:

- Pressure-sensitive mats on the floor at the position of the programmer
- A manual programming device containing an emergency off switch
- Designing the work area of the programmer in such a way that it eliminates unnecessary stress
- Hold-to-run-buttons
- Mandatory reduced speed
- Marking the programming positions
- Planning the programmer's location outside the movement zone (i.e., in a space facing the robot, semiraised on a platform, etc.)
- Turning off safety-related devices with a key only
- Locked turntables during programming

### 6.4.4 Robot Operation and Maintenance Phase

There are many safety-related measures associated with the robotic operation and maintenance phases [4,8,11]. Some of these measures require preventive maintenance regularly and use only the approved parts, develop the necessary safety operations and maintenance procedures, ensure that only authorized and trained personnel operate and maintain robots, report, investigate, and repair any faults or unusual robot motions promptly, ensure the operational readiness of all safety devices (e.g., barriers, interlocks, and guards), block out all concerned power sources during maintenance, observe all government codes and other regulations concerned with the operation and maintenance of robots, provide the initial and periodic training to people associated with robotic operation and maintenance, and make certain that all emergency stops are functional.

## 6.5 Robot-Related Safety Problems Causing Weak Points in Planning, Design, and Operation

There are many weak points in planning, design, and operation, which result in robot-related safety problems in an industrial setting [1,6]. Some of the weak points concerned with planning are as follows:

- *Poor organization of work.* This is an important factor, particularly in programming and stoppages.
- *Improper safety devices.* These comprise improper guards (i.e., too low, containing gaps, or being close to hazard points) and faulty emergency shutdown circuits.
- *Poor spatial arrangement.* This can lead to confusion and the possibility of collision.
- *Unsafe or confused linkages.* These are basically concerned with interfaces between individual machines.

Some of the weak points in design with respect to robot-related safety in industries are shown in Figure 6.2 [12].

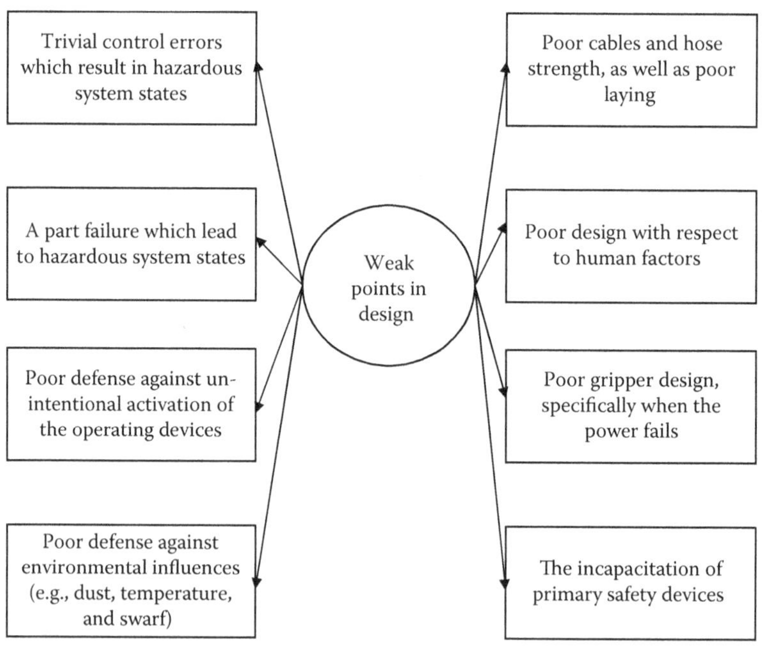


**FIGURE 6.2**
Weak points in design with respect to robot safety.

Finally, some of the weak points in operational procedure with respect to robot-related safety in industrial settings are as follows:

- Permitting counter-safety working procedures during a stoppage
- Poor training to workers who are directly or indirectly concerned with industrial robots
- Failure to give feedback to personnel involved in design and layout regarding weak spots and how to ensure their removal

## 6.6 Robot Safeguard Approaches

There are many robot safeguard approaches [5]. Some of these approaches are described below, separately [5,6].

### 6.6.1 Warning Signs

The warning signs are used in conditions where robots, by virtue of their size, speed, and inability to impart significant amount of force, cannot injure humans. Two examples of such conditions are laboratory and small-part assembly robots. Robots such as these require no special safeguarding as warning signs are quite sufficient for the uninformed humans in the area.

However, it is to be noted that warning signs are quite useful for all robot application areas, irrespective of whether robots possess the ability to injure humans or not.

### 6.6.2 Physical Barriers

Physical barriers are quite useful to safeguard humans in the work areas of robots. However, they are not the ultimate solution to a robot-related safety problem in many cases. The objective of such barriers is to stop humans from reaching over, under, around, or through the barriers into the prohibited robot work zone [1]. Some examples of physical barriers are chain-link fences, safety rails, plastic safety chains, and tagged-rope barriers. Some useful guidelines concerning physical barriers are the following [6]:

- Use safety rails in conditions where projectiles are clearly not a problem at all
- Avoid trapping points within the framework of a barrier by providing a sufficient buffer space between the barrier and the work area
- Chain-link fences and safety rails are very useful in circumstances where intrusion is considered a particular problem

- Make use of fences in circumstances where long-range projectiles are considered a hazard

In any case, when the installation of a peripheral physical barrier is considered, the following questions have to be asked clearly [13]:

- How were perimeter dimensions established?
- Are the perimeter dimensions reliable?
- Can the perimeter be bypassed easily?
- What is being protected?
- Is the protection effective?

### 6.6.3 Intelligent Systems

Intelligent systems make their decisions through remote sensing, hardware, and software. In order to achieve an effective intelligent collision-avoidance system, the robot's operating environment has to be restricted and special sensors and software has to be used widely. This calls for the need for a sophisticated computer for making the right decisions and real-time computations.

Finally, it is to be noted that in most industrial settings, it is generally not possible to restrict the environment.

### 6.6.4 Flashing Lights

Flashing lights are another robot safeguard approach; this approach calls for the installation of a flashing light on the robot itself or at the perimeter of the robot work area. The objective of the flashing light is to alert people in the area that programmed motion is happening or could happen any time.

It is to be noted with care that whenever the flashing lights approach is employed, ensure that the flashing lights approach is employed, ensure that the flashing light is energized continuously during the period when the robot drive power is turned on.

## 6.7 Common Robot Safety Features and Their Functions

A well-designed robot incorporates, to the greatest extent possible, safety features that clearly take into consideration all modes of robot operation (i.e., normal operation, programming, and maintenance) [14,15]. Nonetheless, some of the safety features are generally common to all robots and the others are specific to the robot types. Some of the common robot safety features, along with their corresponding intended functions, given in parentheses, are as follows [15]:

- Power-on button (it energizes all machine power)
- Stop button (it removes control and manipulator power)
- Power disconnect switch (it removes all power at the machine junction box)
- Slow-speed control (it permits program execution at reduced speeds)
- Arm-power-only button (it applies power to the manipulator only)
- Teach pendant trigger (it must be held by the operator for arm power in teach mode)
- Line indicator (it indicates that incoming power is connected at the junction box)
- Hydraulic fuse (it protects against high-speed motion/force in teach mode)
- Manual/automatic dump (it provides means to relieve hydraulic/pneumatic pressure)
- Step button (it permits program execution one step at a time)
- Condition indicators and messages (these provide visual indication by lights or display screens of the condition of the system)
- Remote connections (these permit remote control of essential machine/safety functions)
- Parity checks, error detecting, and so on (whereby the computer approaches for self-checking a variety of functions)
- Hardware stops (absolute control on travel/movement limits)
- Program reset (this drops the system out of playback mode)
- Hold/run button (it stops arm motion, but leaves power on)
- Control-power-only button (it applies power to the control section only)
- Software stops (computer-controlled travel limit)
- Teach/playback mode selector (it provides the operator with control over the operating mode)
- Servo-motor brake (it maintains the arm position at a standstill)

## 6.8 Safety Considerations for Robotized Welding Operations

Today, in industries, robots are often used to perform various types of welding operations. Welding robots undertake welding processes such as gas tungsten arc, resistance (spot), gas metal arc, and laser [6,16].

There are many factors that have to be considered with care for the safe operation of welding robots. Twelve of these factors are shown in Figure 6.3 [6,16]. The factor "torch motion simulation" is concerned with the simulation of the welding torch motion prior to striking the arc. The factor "common ground" calls for not sharing the common ground in situations where a welding robot is joining conventional welding systems with high-frequency power supplies.

The factor "area contamination" is concerned with ensuring that welding robots do not contaminate areas where spatter or harmful light rays may generate adverse effects.

The factor "environment" is concerned with finding an area that is free from dust, high humidity, vibration, oil, and smoke and also has low traffic.

The factor "power switches" is concerned with locking such switches to the control panel and welding power supply. The factor "interlocking switches" is concerned with installing such switches to the welding jigs and fixtures, which indicate safe clamping.

The factor "robot floor" is concerned with choosing a suitable floor, surrounding and directly beneath the robot. In this case, it is to be noted that

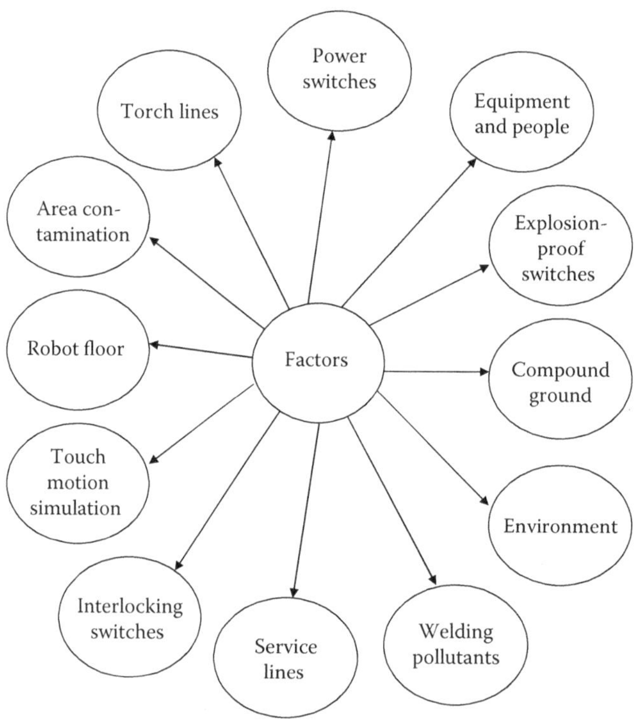


**FIGURE 6.3**
Factors to be considered for the safe operation of welder robots.

metal plates and explosion-proof concrete offer a good solution for problems caused by sparks and hot metal falling to the floor. The factor "equipment and people" is concerned with preventing damage and injury to equipment and people, respectively. In this case, it is to be noted that by placing spatter shields and ultraviolet light-filtering screens around the welding table, the positioner and workstation can prevent damage and injury to equipment and people.

The factor "service lines" includes electricity, water, air, exhaust, and ventilation lines. Here, the aim should be to locate such service lines underneath the floor or in floor channels away from the welding torch and torch lines, in addition to locating them away from the robot's work area. Furthermore, avoid having water and air lines in the same channel. The factor "torch lines" is concerned with avoiding the entanglement of such lines with the joints of the robot during extreme movement. The items such as pivots, strings, pulleys, and counterweights can be used to eliminate this problem.

The factor "welding pollutants" is concerned with alleviating problems of welding pollutants. In this case, an electrostatic precipitator may help to reduce pollutants as well as the use of a centrally located charcoal collector to reduce welding gases. It is to be noted that this collector draws the residual gases through its torch attachment. Thus, it is very useful in circumstances where it is not possible to have an underfloor exhaust system as well as where the robot has to perform many welds over fairly long distances.

Finally, the factor "explosion-proof switches" is concerned with using such switches whenever possible.

As there are specific hazards and safeguards associated with robotized laser welding, gas-shielded arc welding, and resistance welding, each of these topics is discussed below, separately [6,16].

### 6.8.1 Robotized Laser Welding Hazards and Safety Measures

Laser welding is the type of welding where energy is generated by a laser beam and focused at a point. The hazards that emanate from the use of robotized laser welding are those connected with fumes and electromagnetic radiation. Welding lasers can cause fire and hazards to both eyes and skin (diffuse reflections).

When the robot is in its normal operation mode, there are many safety measures associated with power lasers. Some of these safety measures are as follows [6,16]:

- *Safety measure I: Emission indicator.* It indicates that the laser is in its active mode.
- *Safety measure II: Beam attenuator.* It helps to stop light from the laser entering the enclosure while humans are still inside.

- *Safety measure III: Enclosure.* It helps to stop exposure to laser light.
- *Safety measure IV: Remote interlock.* It is used for stopping access to the power laser enclosure during laser action.
- *Safety measure V: Warning signs.* These are to be placed at accesses to the enclosure for their effectiveness.
- *Safety measure VI: Key control.* It is useful for stopping unauthorized use.
- *Safety measure VII: Training.* It is very helpful for familiarization with the product and for hazard control.

### 6.8.2 Hazards and Safety Measures for Robotized Gas-Shielded Arc Welding

Gas-shielded arc welding is the type of welding where both the arc and molten pool are shielded from the atmosphere by a gas [16]. When this type of welding is robotized, there are many types of associated hazards. Some of these hazards are resulting fumes, hot metal sparks, electric shock from the torch, fire from hot metal welded parts, electromagnetic radiation, mechanical hazards, and program corruption [6,16]. For modes normal, programming, and maintenance solutions to these hazards are available in [16].

For the normal mode of robotized operations, selected hazards associated with gas-shielded arc welding and their corresponding safety measures (given in parentheses) are as follows [6]:

- Phosgene from chlorinated solvents on work items (dry all parts with care after any cleaning process before welding)
- Ozone from arc (provide appropriate fume extraction fans in the general area of work)
- Electric shock from torch (place torch inside the work perimeter)
- Fire from hot metal sparks (ensure that no flammable materials exist inside the perimeter guard and that all cables and pipes within the perimeter guard framework are properly protected)
- Fire from hot metal welded parts (ensure that the unloaded items are properly kept away from flammable materials)
- Sparks from hot metal (position an interior perimeter guard and carefully ensure that there is no flammable material inside the guarded workspace. Also, keep all pipes and cables inside the guarded workspace properly protected)
- Smoke from leftover oil on work items (minimize oil on the work surface to decrease oil fumes as well as to ensure good welds)
- Magnetic radiation (minimize this radiation to a normal level; computer enclosures should furnish an adequate shield to electronics)

### 6.8.3 Hazards and Safety Measures for Robotized Resistance Welding

Resistance welding is the type of welding in which, at some point during the process, force is put to the surfaces in contact, and in which welding heat is produced by sending electric current through the resistance at, and adjacent to, these surfaces [6,16]. The use of robots is quite useful to minimize some of the hazards that are traditionally associated with resistance welding.

These associated hazards may be categorized into the following three groups [6]:

- *Electrical hazards.* These hazards are magnetic fields, high/medium voltage, and access to welding control equipment. The safety measures applicable to the normal mode of robot operation are the identification of workers wearing pacemakers, placing primary supply and welding transformers in a safe location, and locking to stop unauthorized access.
- *Mechanical hazards.* Trapping by work-handling equipment and shocks from low-voltage conductors are the two examples of mechanical hazards. The safety measures applicable to the normal mode of robotized operation are providing satisfactory guarding/interlocks to stop access to the hazard zone and balance support for minimizing mechanical stress on the robot arm.
- *Other hazards.* These hazards are fumes, noise, and fire. Some of the safety measures applicable to the normal mode of robotized operation are placing extractor fans in an "on" position, keeping closed containers of weld sealers, and silencers on the air exhaust.

## 6.9 Problems

1. What are the unique robot safety problems? List at least six such problems.
2. What are the basic robot hazards? Discuss each of these hazards in detail.
3. Discuss the roles of robot manufacturers and users with regard to safety.
4. Discuss safety considerations in robot design and installation phases.
5. Discuss safety considerations in robot programming and operations and maintenance phases.
6. List at least eight weak points in design with respect to robot-related safety.

7. Describe the following two robot safeguard approaches:
   a. Warning signs
   b. Flashing lights
8. List at least 10 common robot-related safety features along with their functions.
9. What are the factors to be considered for the safe operation of welder robots?
10. What are the robotized laser welding hazards and safety measures?

## References


1. American National Standards Institute, *American National Standard for Industrial Robots and Robot Systems—Safety Requirements, ANSI/RIA R15.06-1986*, New York, 1986.
2. Japanese Industrial Safety and Health Association, *An Interpretation of the Technical Guidance on Safety Standards in the Use, Etc., of Industrial Robots*, Tokyo, 1985.
3. Ziskovsky, J.P., Risk analysis and the $R^3$ factor, *Proceedings of the Robots 8th Conference*, Vol. 2, June 1984, pp. 15.9–15.21.
4. Ziskovsky, J.P., Working safely with industrial robots, *Plant Engineering*, May 1984, pp. 81–85.
5. Addison, J.H., *Robotic Safety Systems and Methods: Savannah River Site, Report No. DPST-84-907 (DE 35-008261)*, December 1984, issued by E.I. du Pont de Nemours and Company, Savannah River Laboratory, Aiken, South Carolina 29808.
6. Dhillon, B.S., *Robot Reliability and Safety*, Springer-Verlag, New York, 1991.
7. Ramachandran, V., Vajpayee, S., Safety in robotic installations, *Robotics and Computer-Integrated Manufacturing*, Vol. 3, 1987, pp. 301–309.
8. Russell, J.W., *Robot Safety Considerations: A Checklist, Professional Safety*, December 1983, pp. 36–37.
9. Bellino, J.P., Meagher, J., Design for safeguarding, *Proceedings of the Robots East Seminar*, Boston, Massachusetts, October 1985, pp. 24–37.
10. Nicolaisen, P., Ways of improving industrial safety for the programming of industrial robots, *Proceedings of the 3rd International Conference on Human Factors in Manufacturing*, November 1986, pp. 263–276.
11. Jiang, B.C., Robot safety: Users' guidelines, *Trends in Ergonomics/Human Factors III*, ed. W. Karwowski, Elsevier, Amsterdam, 1986, pp. 1041–1049.
12. Akeel, H.A., Intrinsic robot safety, *Working Safely with Industrial Robots*, ed. P.M. Strubhar, Robotics International of the Society of Manufacturing Engineers, Publications Development Department, One SME Drive, P.O. Box 930, Dearborn, Michigan, 1986, pp. 61–68.

13. Marton, T., Pulaski, J.L., Assessment and development of HF related safety designs for industrial robots and robotic systems, *Proceedings of the Human Factors Society 31st Annual Meeting*, 1987, pp. 176–180.
14. Bararett, R.J., Bell, R., Hudson, P.H., Planning for robot installation and maintenance: A safety framework, *Proceedings of the 4th British Robot Association Annual Conference*, 1981, pp. 18–21.
15. Clark, D.R., Lehto, M.R., Reliability, maintenance, and safety of robots, *Handbook of Industrial Robotics*, ed. S.Y. Nof, John Wiley and Sons, New York, 1999, pp. 717–753.
16. *Safeguarding Industrial Robots, Part II: Welding and Allied Processes*, The Machine Tool Trades Association (MTTA), London, 1985.

# 7

# Robot Accidents and Analysis

## 7.1 Introduction

An accident is an unplanned and undesired event; every year, thousands of lives are lost due to work-related injuries. For example, in 2007, in the United States, 5488 workers died from job-related injuries and about four million workers suffered from nonfatal work-related injuries or illnesses [1,2].

Needless to say, accidents are an important issue and, in regard to robots, there have been many fatal and nonfatal accidents over the years. For example, as per [3], up to the early 1980s, there were five fatal accidents involving robots: four in Japan and one in the United States; the up-to-date information on fatal and nonfatal robot accidents in the United States is available from [4].

This clearly indicates that robot-related accidents are pressing problems and thus require careful attention from industry. This chapter presents various important aspects of robot accidents and analysis.

## 7.2 Some Examples of Robot-Related Accidents

Over the years, many robot-related accidents have occurred. Some examples of such accidents are as follows [3,5–9]:

- A worker entered the robot cell to clean its sensors and was killed because he overlooked stated lockout procedures.
- A repair person climbed over a safety fence around the robot work area without turning off power to the robot and performed tasks in the area while the robot was temporarily stopped. When the robot recommenced movement, it pushed the repair person into a nearby grinding equipment and, consequently, the person died.
- A robot operator went to check or fix a robot failure without locking the robot. During the repair process, the operator activated the robot

and the arm of the robot crushed the operator against a part being transported on a conveyor and the operator died.

- A worker stepped between a machine (a planer) and the robot it was servicing, and turned off the circuit that was emitting activating signals from the planer to the robot. After carrying out the required task, the worker switched on the same circuit while still being in the workspace area of the robot. The robot started its operation and crushed the worker to death against the planer.
- A worker was freeing a jam at the end of an oven and became caught between the conveyor belt of the oven and the robotic arm and got killed.
- A worker turned on a welding robot while another worker was still in the robot's work area. The robot pushed the worker in its work area into the positioning fixture; the worker was consequently killed.
- A worker entered the caged robotic palletizer cell while the robotic palletizer was still running. The worker's torso was crushed by the arms of the robotic palletizer as it tried to pick up boxes on the roller conveyor and the worker died.
- A worker climbed onto a conveyor belt in motion for recovering a faulty part when the robot serving the line was stopped on a program point temporarily. The robot crushed the worker to death when the operation was restarted.
- A worker was troubleshooting a robotic arm used for removing CD jewel cases from an injection molding machine, when the arm cycled and struck the worker, who died two weeks later.
- A worker violated safety devices for entering a work cell while the material handling robot was operating it in automatic mode. The worker was trapped between the robot and a post anchored to the floor. The injured worker died a few days later.
- A worker freeing a jam at the end of an oven was caught between a robotic arm and the conveyor belt of the oven and was killed.
- A worker holding the robot controls in his hand activated the robot while bending over the wheel to check the settings. The robot pinned the worker against the wheel and crushed him to death.
- A repair person observing the operation of a robot entered the work area of another robot without being aware of it. When the other robot moved back to its home location, it knocked down the repair person who suffered a cervical strain as the result of being knocked down.
- During the manual operation, a robot's arm went out of sequence; consequently, when the robot operator tried to regulate it, the operator received a cut on the head.

- A worker assisting an electrical engineer in troubleshooting a malfunctioning robot was crushed between the lifting arm and the lower frame of the robot when the lifting arm dropped.

## 7.3 Robot Accidents: Causes and Sources

Over the years, many studies have been conducted to determine the causes for the occurrence of robot-related accidents. The findings of two studies are presented below.

### 7.3.1 Study I

A Japanese study based on a survey of robots reported the following causes (with corresponding percentages in parentheses) for 18 near-accidents [10]:

- Erroneous robot movement during testing or teaching (16.6%)
- Wrong action by the robot during manual operation (16.6%)
- Wrong movement of peripheral equipment during normal operation (5.6%)
- Wrong movement during repair, regulation, and checking (16.6%)
- Erroneous robot movement during normal operation (5.6%)
- Sudden entry of the human into the robot work area (11.2%)
- Wrong movement of peripheral equipment during testing or teaching (16.6%)
- Others (11.2%)

The above percentages, given in parentheses, indicate that robot accidents are most likely during setting, adjusting, or other manual operations. Some of the possible reasons for the occurrence of these accidents are as follows [5]:

- Workers frequently forget about hazards associated with robot systems under normal or abnormal conditions.
- Workers involved with robot systems become preoccupied and careless/complacent.
- Workers involved with robot systems often take chances as opposed to following the stated procedures carefully/vigilantly.

### 7.3.2 Study II

A General Motors Corporation study reported the following causes for many robot incidents [5]:

- The presence of an ignorant individual in the operating enclosure of the robot—people with authorization ignorant of the ramifications of the robot program
- Lack of vigilance from involved workers to adjacent robots

There are many possible sources for robot-related accidents, which may be divided into two groups as follows [5,11,12]:

- *Group I: Engineering factors.* This group includes factors such as failure of the control panel or peripheral equipment, poor software design, high speed of the robotic arm, the robot worker work environment, problems associated with the design of the gripper and/or the control panel, robot system failures (electrical, mechanical, etc.), inadvertent movement of the robot, unintentional contact with the start and other switches, and failure of safety devices.
- *Group II: Behavioral and organizational factors.* This group includes factors such as wrong procedures for initial start-up of the robot, programmed movement of the robot unknown to the robot operator, poor robot-related training programs for people such as repair persons, operators, programmers, and supervisory personnel, lack of awareness among operators of the work function of the robot or an operator being unaware that another person is within the danger space, and curiosity leading a nonoperating person to approach a robot.

It is to be noted that some of the above engineering, behavioral, and organizational factors cannot be strictly divided into one group or another. Furthermore, there could be, direct or indirect, overlap between various causes and sources for robot-related accidents. The main objective here is to highlight the direction that a safety professional should follow in further analyzing the underlying sources and causes of robot-related accidents.

## 7.4 Effects of Robot-Related Accidents

The effects of robot-related accidents on humans may range from no injury to a fatality. A study of 32 robot-related accidents grouped the effects of the accidents in the following two categories [13]:

- *Category I*: Pinch-point (i.e., a person or part of his/her body pinched between two robot parts or between the robot itself and some external item)
- *Category II*: Impact

The percentage breakdowns between these categories (i.e., pinch-point and impact) were roughly 56% and 44%, respectively. Some examples of the effects of robot-related accidents are as follows [5]:

- Pinch-point injury to person's chest
- Pinch-point injury to person's hand
- The arm of the robot strikes a person's head
- The arm of the robot programmer impacted by the robotic arm
- The arm of the robot strikes a person

The following workers were involved in the 32 robot accidents [5,13,14]:

- Line operators (23 times)
- Maintenance personnel (six times)
- Programmers (three times)

The breakdowns between impact and pinch among these workers were as follows [5,13]:

- *Impact*: Two (programmers), two (maintenance personnel), 10 (line operators)
- *Pinch*: One (programmers), four (maintenance personnel), 13 (line operators)

Finally, it is to be noted from the above data that the line operators were involved around 72% of the robot-related accidents.

## 7.5 Robot-Related Accidents at Manufacturer and User Facilities

Robot-related accidents can occur either during the development process at the manufacturer's facilities or during the commissioning period, and/or in use at the premises of the user [5,15]. Usually, accidents at the manufacturing facilities occur during the test run period or during the programming period. The main reason for the occurrence of accidents during the robot programming process is that people frequently stand within the work area of the robot.

The opportunities for the occurrence of a robot-related accident are quite high during the test run process, since there is no way to be absolutely sure that the program is totally bug-free or that the associated hardware is wholly reliable.

The likelihood of robot-related accidents at the facilities of the user is higher than at the manufacturer's facilities basically due to the following two reasons:

- The length of the robotic operation is much higher.
- The surroundings in which the robots carry out their assigned mission are subject to intrusion from people who may not possess proper knowledge of the safety needs of robots.

At the users' facilities, robot-related accidents may occur during programming, maintenance, or when robots carry out operations in the automatic mode. Normally, to mend robots, the programmers enter the work area of robots and this creates hazardous situations. During maintenance of a robot, interactions between the maintenance personnel and the robot are quite close, and any carelessness in safety precautions can result in an accident.

During the operations of robots in the automatic mode an accident can occur more as a result of human error than anything else, because intrusions inside work zones of robots in the automatic mode can be very risky. Additional information, directly or indirectly, concerned with robot-related accidents at user facilities is available in the preceding sections of this chapter.

## 7.6 Useful Recommendations to Prevent Human Injury by Robots

Over the years, many organizations around the world have proposed useful recommendations to prevent injury to humans by robots. The recommendations proposed by the National Institute for Occupational Safety and Health in the United States were directed at the following three areas [16]:

- Robot system design
- Supervision of workers
- Training of workers

The recommendations concerning each of the above three areas are presented below, separately.

### 7.6.1 Robot System Design

The following recommendations are concerned with both new designs and existing robotic equipment:

- Ensure that control and operational areas of the robot have appropriate levels of illumination.
- Ensure that adequate clearance distances are provided around all moving parts of the robot system.
- Ensure that appropriate remote "diagnostic" instrumentation is incorporated so that troubleshooting of the system can be performed mainly from places outside the work area of the robot.
- Ensure that proper physical barriers incorporating gates with electrical interlocks are provided so that, at the moment of opening the gate, the operation of the robot is stopped instantly.
- Ensure that floors or working surfaces contain adequate visible marks that show the work area of the robot clearly.
- Ensure that proper barriers between freestanding objects and the robot equipment are provided appropriately.
- Ensure that, as a back-up to motion sensors, floor sensors, light curtains, or electrical interlocks (that turns off the robot instantly in situations whenever the barrier is crossed by a person) are included.

### 7.6.2 Supervision of Workers

The following two recommendations are for the work supervisors:

- Recognize that, with time, experienced personnel carrying out automated tasks may become overconfident, complacent, or inattentive to the danger present in automated items.
- Ensure that no human enters the work area of the robot prior to putting the robot on "hold," or on a reduced speed mode.

### 7.6.3 Training of Workers

The following recommendations are concerned with training of workers:

- Ensure that training clearly emphasizes that workers involved with robots must be knowledgeable about all working aspects of the robot, prior to carrying out their assigned tasks at robot workstations.
- Provide appropriate training, specific to the particular robot under consideration, to operation personnel, programmers, and maintenance personnel.

- Ensure that training clearly emphasizes that all robot operators must never be within reach of operating robots.
- Ensure that training clearly emphasizes the operating of robots at reduced speeds—consistent with a satisfactory human response time to avoid risks during the programming process—by programmers, operators, and maintenance personnel and that they should be alert to each and every possible pinch-point within the robots' operational areas. Two typical examples of pinch-points are walls and poles.
- Provide appropriate refresher training courses to experienced operators, maintenance personnel, and programmers that emphasize new technological developments and robot-related safety.

## 7.7 Methods for Performing Robot Accident Analysis

There are many methods for analyzing robot-related accidents. Some of these methods are presented below.

### 7.7.1 Estimating Probability of an Accident Occurrence Related to the Operation of a Robot

This section presents an expression that can be used to estimate the probability of an accident occurring related to the operation of a robot. The probability of such an occurrence is expressed by [5]

$$P_{ar} = P_A + P_B + P_C \tag{7.1}$$

where

$P_{ar}$ is the probability of an accident occurrence from the operation of a robot.

$$P_A = \sum \{P_1 + P_2(1 - P_{11})\}\{1 + P_3 + P_4\} \tag{7.2}$$

$$P_B = \sum \{P_5 + P_6(1 - P_{11})\}\{1 + P_7 + P_8\} \tag{7.3}$$

$$P_c = \sum (P_9 + P_{10})(1 - P_{11}) \tag{7.4}$$

The symbols used in the above three equations (i.e., Equations 7.2 through 7.4) are defined below:

$P_1$ is the probability of the occurrence of an irreversible human failure that could result in or allow an accident.

$P_2$ is the probability of the occurrence of a reversible human failure that could result in or permit a mishap.

$P_3$ is the probability of the robot possessing an adverse characteristic that could result in human failure (error).

$P_4$ is the probability of the robot experiencing an adverse environmental condition that could result in human failure (error).

$P_5$ is the probability of occurrence of the failures that could result in mishaps (in this case no corrective measure is possible).

$P_6$ is the probability of occurrence of those malfunctions that could result in accidents unless the appropriate actions are taken in a timely manner.

$P_7$ is the probability of the robot having an adverse characteristic that could result in human failure (error).

$P_8$ is the probability of the robot experiencing an adverse environmental condition that could result in robot failure.

$P_9$ is the probability of the robot possessing an adverse characteristic that could result in injury, damage, or loss in the absence of material failure or error.

$P_{10}$ is the probability of the robot experiencing an adverse environmental condition that could lead to damage or injury in the absence of error or failure.

$P_{11}$ is the probability of necessary action taken as required.

### 7.7.2 Root Cause Analysis

Root cause analysis (RCA) may simply be described as a systematic investigation approach that uses information collected during an assessment of an accident for determining the underlying factors for the deficiencies that resulted in the accident [17]. The method was originally developed by the U.S. Department of Energy to investigate industrial accidents [18,19].

The performance of the RCA helps in better understanding the causal factors in the sequence of evolving events; the RCA process is concluded with recommendations for improvements on the basis of the investigational findings [19,20].

The steps generally involved to perform RCA are as follows [20,21]:

- Educate everyone involved in RCA
- Inform appropriate staff members when the occurrence of a sentinel event is reported
- Form an RCA team by including appropriate individuals
- Prepare for and hold the first team meeting
- Determine the event sequence
- Separate and identify each event sequence that may have been a contributory factor in the sentinel event occurrence
- Brainstorm about the factors surrounding the selected events that may have, directly or indirectly, contributed to the occurrence of the sentinel event
- Affinitize with the brainstorming session results
- Develop the appropriate action plan
- Distribute the RCA document and the associated action plan to all individuals concerned

#### *7.7.2.1 RCA Software*

Over the years, various types of software packages have been developed for conducting RCA. The benefits cited for the automation of RCA include better data organization, enhanced follow-up capabilities, reduction in analysis time, improved rigor, and easier reporting capabilities [17]. Some of the available RCA software packages in the market are as follows [17]:

- BRAVO, developed by JBF Associates, Inc., 1000 Technology Drive, Knoxville, TN 37939, USA.
- REASON, developed by Decision Systems, Inc., 802 N. High St., Suite C, Longview, TX 75601, USA.
- PROACT, developed by Reliability Center, Inc., P.O. Box 1421, Hopewell, VA 23860, USA.
- TAPROOT, developed by System Improvements Inc., 238 S. Peters Road, Suite 301, Knoxville, TN 37923-5224, USA.

#### *7.7.2.2 RCA Advantages and Disadvantages*

Today, RCA is often used to investigate various types of industrial accidents. Some of its advantages are as follows [22]:

- It is an effective tool to address and identify organizational and systems issues.

- It is a well-structured and process-focused method.
- The systematic application of this method can uncover common root causes that link a disparate collection of accidents.

In contrast, some of the disadvantages of RCA are as follows [22]:

- It is quite a time-consuming and labor-intensive method.
- It is quite possible to be tainted by the bias of hindsight.
- It is impossible to determine precisely if the root cause established by the analysis is really the actual cause for the occurrence of the accident.
- In essence, RCA is an uncontrolled case study.

### 7.7.3 Fault Tree Analysis

The fault tree method was developed in the early 1960s to analyze the Minuteman Launch Control System with respect to safety at the Bell Telephone Laboratories [23]. Today, it is widely used throughout the world to perform various types of studies on reliability and safety. The method is described in Chapter 4.

Here, the application of the fault tree method to perform robot accident analysis is demonstrated through the following two examples:

**EXAMPLE 7.1**

Assume that at a robotic facility, a robot-related accident involving a human, caused by sudden robot movement, can occur due to the occurrence of the following: human in the robot's work area and sudden robot movement. The event "human in robot work area" can be either intentional entry or unintentional entry. The event "sudden robot movement" can occur due to the occurrence of the following: power supply on and unexpected start signal.

In turn, the event "power supply on" can occur either due to control mechanism malfunction or human error. With the aid of the fault tree symbols given in Chapter 4, develop a fault tree for the top event "robot-related accident, involving a human, caused by sudden robot movement."

A fault tree for the example is shown in Figure 7.1. The single capital letters in the figure denote corresponding fault events (e.g., *A*: unexpected start signal, *Y*: Human in robot work area, and *Z*: Power supply on).

**EXAMPLE 7.2**

Assume that in Figure 7.1, the probabilities of occurrence of fault events *A*, *B*, *C*, *D*, and *E* are 0.05, 0.1, 0.04, 0.15, and 0.06, respectively. With the aid of Chapter 4, calculate the probability of occurrence of the top

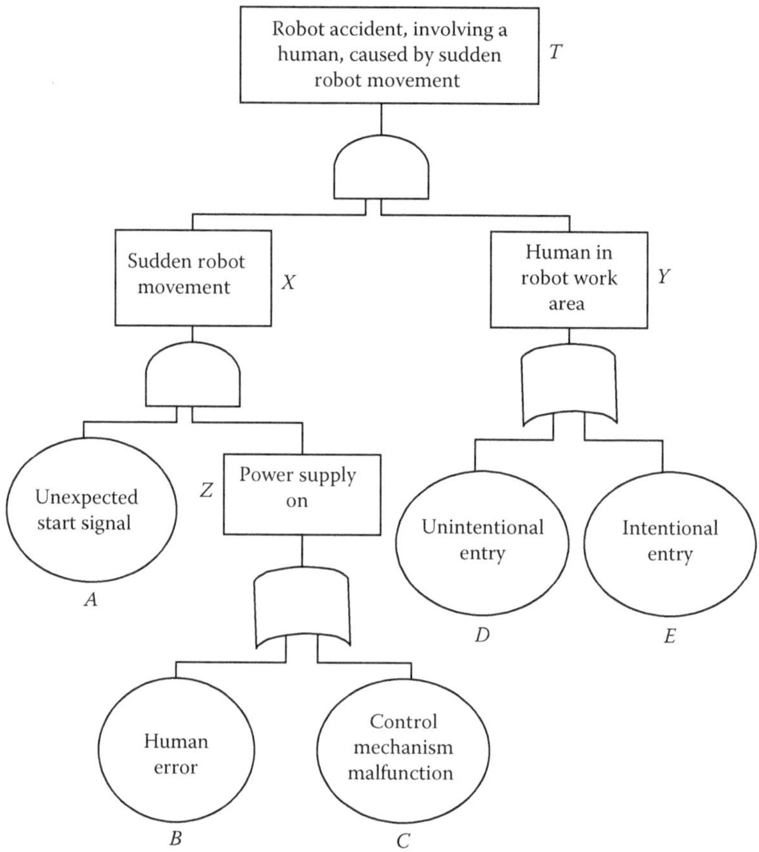


**FIGURE 7.1**
A fault tree for the top event: Robot-related accident, involving a human, caused by sudden robot movement.

event *T*: Robot-related accident, involving a human, caused by sudden robot movement. With the aid of material presented in Chapter 4, we calculate the probabilities of occurrence of fault events *Z*, *Y*, *X*, and *T* as follows:

The probability of the occurrence of event *Z* is

$$\begin{aligned} P(Z) &= 1 - (P(B))(1 - P(C)) \\ &= 1 - (1 - 0.1)(1 - 0.04) \\ &= 0.136 \end{aligned} \tag{7.5}$$

where

$P(B)$ is the occurrence probability of event *B*.
$P(C)$ is the occurrence probability of event *C*.

Similarly, the probability of the occurrence of the event $Y$ is

$$\begin{aligned} P(Y) &= 1 - (1 - P(D))(1 - P(E)) \\ &= 1 - (1 - 0.15)(1 - 0.06) \\ &= 0.201 \end{aligned} \tag{7.6}$$

where
$P(D)$ is the occurrence probability of event $D$.
$P(E)$ is the occurrence probability of event $E$.

The probability of the occurrence of the intermediate event $X$ is

$$\begin{aligned} P(X) &= P(A)P(Z) \\ &= (0.05)(0.136) \\ &= 0.0068 \end{aligned} \tag{7.7}$$

where
$P(A)$ is the occurrence probability of event $A$.

The top event $T$ (robot-related accident, involving a human, caused by sudden robot movement) probability of occurrence is

$$\begin{aligned} P(T) &= P(X)P(Y) \\ &= (0.0068)(0.201) \\ &\cong 0.0014 \end{aligned} \tag{7.8}$$

Thus, the probability of the occurrence of the top event $T$ (robot-related accident, involving a human, caused by sudden robot movement) is approximately 0.0014. The Figure 7.1 fault tree with calculated and given probability values is shown in Figure 7.2.

### 7.7.4 Markov Method

The Markov method is widely used to perform various types of reliability and availability analyses and is described in Chapter 4. Here, the application of the Markov method to perform robot-related accident analysis is demonstrated through the following two examples:

**EXAMPLE 7.3**

Assume that a robot system can fail safely or with an accident. The robotic system's safe failure rate is $\lambda_s$ and its failing with an accident failure rate is $\lambda_a$. The robot system state-space diagram is shown in Figure 7.3. The numerals in the diagram circle and boxes denote the robot system states. Develop expressions for the robot system state probabilities and mean

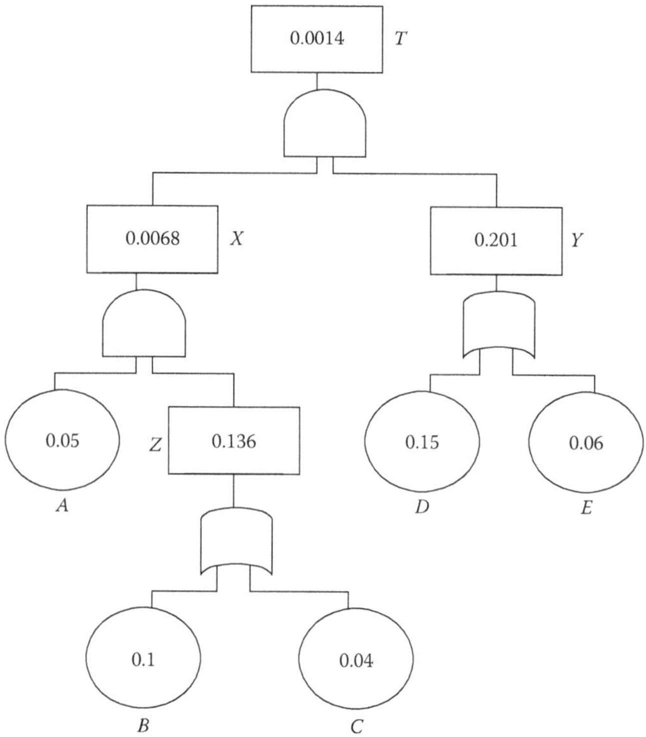


**FIGURE 7.2**
Redrawn Figure 7.1 fault tree with calculated and given event occurrence probability values.

time to failure by using the Markov method and assuming that robot system failures occur independently and its failure rates are constant.

The following symbols are associated with the Figure 7.3 state-space diagram and its associated equations:

$j$ is the state of the robot system: $j = 0$ means the robot system is operating normally; $j = 1$ means the robot system has failed safely; $j = 2$ means the robot system has failed with an accident.
$t$ is time.

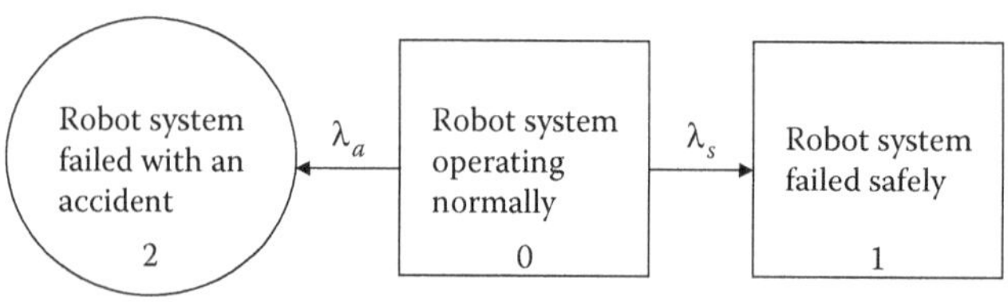


**FIGURE 7.3**
Robot system state-space diagram.

$P_j(t)$ is the probability that the robotic system is in state $j$ at time $t$, for $j = 0,1,2$.
$\lambda_s$ is the constant robot system, failing safely, failure rate.
$\lambda_a$ is the constant robot system, failing with an accident, failure rate.

By using the Markov method presented in Chapter 4, we write down the following set of differential equations for the Figure 7.3 diagram:

$$\frac{dP_0(t)}{dt} + P_0(t)(\lambda_s + \lambda_a) = 0 \tag{7.9}$$

$$\frac{dP_1(t)}{dt} - P_0(t)\lambda_s = 0 \tag{7.10}$$

$$\frac{dP_2(t)}{dt} - P_0(t)\lambda_a = 0 \tag{7.11}$$

At time $t = 0$, $P_0(0) = 1$, $P_1(0) = 0$, and $P_2(0) = 0$.
By solving Equations 7.9 through 7.11, we get

$$P_0(t) = e^{-(\lambda_s + \lambda_a)t} \tag{7.12}$$

$$P_1(t) = \frac{\lambda_s}{\lambda_s + \lambda_a}\left[1 - e^{(\lambda_s + \lambda_a)t}\right] \tag{7.13}$$

$$P_2(t) = \frac{\lambda_a}{\lambda_s + \lambda_a}\left[1 - e^{-(\lambda_s + \lambda_a)t}\right] \tag{7.14}$$

By integrating Equation 7.12 over the interval [0,∞], we get the following expression for the robot system mean time to failure [24]:

$$\begin{aligned} MTTF_{rs} &= \int_0^\infty e^{-(\lambda_s + \lambda_a)t}\,dt \\ &= \frac{1}{\lambda_s + \lambda_a} \end{aligned} \tag{7.15}$$

where
$MTTF_{rs}$ is the robot system mean time to failure.

**EXAMPLE 7.4**

Assume that the robotic system's failing with an accident failure rate is 0.0002 failures per hour and its failing safely failure rate is 0.0005 failures per hour. Calculate the robot system failure rate failing with an accident probability during a 500-h mission and its mean time to failure.

By substituting the specified data values into Equation 7.14, we obtain

$$P_2(500) = \frac{0.0002}{0.0005 + 0.0002}\left[1 - e^{-(0.0005+0.0002)(500)}\right]$$
$$= 0.0843$$

Similarly, by inserting the specified data values into Equation 7.15, we get

$$MTTF_{rs} = \frac{1}{0.0005 + 0.0002}$$
$$= 1428.57 \text{ h}$$

Thus, the robotic system failing with an accident probability and mean time to failure are 0.0843 and 1428.57 h, respectively.

## 7.8 Problems

1. Discuss the causes for the occurrence of robot-related accidents.
2. What are the sources of robot-related accidents?
3. Discuss the effects of robot-related accidents with regard to humans.
4. Discuss robot-related accidents at the robot manufacturer's facilities.
5. Discuss robot-related accidents at the robot user's facilities.
6. Discuss recommendations to prevent human injury by robots.
7. What are the general steps involved to perform root cause analysis?
8. What are the advantages and disadvantages of root cause analysis?
9. Prove Equations 7.12 through 7.14 with the aid of Equations 7.9 through 7.11.
10. Assume that a robot system can fail safely or with an accident and its failing with an accident failure rate is 0.0004 failures per hour and its failure rate in failing safely is 0.0009 failures per hour. Calculate the probability for the robot system failing safely during a 200-h mission and its mean time to failure.

## References

1. Steenland, K., Burnett, C., Lalich, N., Ward, E., Hurrall, J., Dying for work: The magnitude of U.S. mortality from selected causes of death associated with occupation, *American Journal of Industrial Medicine*, Vol. 43, 2003, pp. 461–482.

2. Bureau of Labor Statistics, United States Department of Labor, *Workplace Injuries and Illnesses in 2007*, Washington, DC, 2008.
3. Altamuro, V.M., *Working Safely with the Iron Collar Worker*, National Safety News, July 1983, pp. 38–40.
4. United States Department of Labor, Occupational Safety and Health Administration (OSHA), *Reports No. 0626300, 0950636, 0728900, 0627100, 0111700, 0454712, and 0552652*, 200 Constitution Avenue, Washington, DC.
5. Dhillon, B.S., *Robot Reliability and Safety*, Springer-Verlag, New York, 1991.
6. United States Department of Labor, Occupational Safety and Health Administration (OSHA), *Report No. 0552652*, Washington, DC, October 10, 2006.
7. Nicolaisen, P., Safety problems related to robots, *Robotics*, Vol. 3, 1987, pp. 205–211.
8. Lauch, K.E., New standards for industrial robot safety, *CIM Review*, Spring 1986, pp. 60–68.
9. Japanese Ministry of Labor, Tokyo, *Study on Accidents Involving Industrial Robots, Report No. PB83239822*, 1982; available from the National Technical Information Service (NTIS), Springfield, Virginia 22161.
10. Sugimoto, N., Safety engineering on industrial robots and their draft standard safety requirements, *Proceedings of the 7th International Symposium on Industrial Robots*, October 1977, pp. 461–468.
11. Rahimi, M., System safety approach to robot safety, *Proceedings of the Human Factors Society's 28th Annual Meeting*, 1984, pp. 102–106.
12. Rahimi, M., Systems safety for robots: Energy barrier analysis, *Journal of Occupational Accidents*, Vol. 8, 1986, pp. 127–138.
13. Jiang, B.C., Gainer, C.A., A cause and effect analysis of robot accidents, *Journal of Occupational Accidents*, Vol. 9, 1987, pp. 27–45.
14. Hirschfeld, R.A., Aghazadeh, F., Chapleski, R.C., Survey of robot safety in industry, *The International Journal of Human Factors in Manufacturing*, Vol. 3, No. 4, 1993, pp. 369–379.
15. Ramachandran, V., Vajpayee, S., Safety in robotic installations, *Robotics and Computer-Integrated Manufacturing*, Vol. 3, 1987, pp. 301–309.
16. National Institute for Occupational Safety and Health, *Request for Assistance in Preventing the Injury of Workers by Robots, Report No. PB85-236818*, Cincinnati, Ohio, December 1984; available from the National Technical Information Service (NTIS), Springfield, Virginia.
17. Latino, R.J., Automating root cause analysis, in *Error Reduction in Health Care*, ed. P.L. Spath, John Wiley and Sons New York, 2000, pp. 155–164.
18. Lement, B.S., Ferrera, J.J., Accident causation analysis by technical experts, *Journal of Product Liability*, Vol. 5, No. 2, 1982, pp. 145–160.
19. Dhillon, B.S., *Safety and Human Error in Engineering Systems*, CRC Press, Boca Raton, Florida, 2013.
20. Dhillon, B.S., *Human Reliability and Error in Medical System*, World Scientific Publishing, River Edge, New Jersey, 2003.
21. Burke, A., *Root Cause Analysis, Report*, 2000; available from the Wild Iris Medical Education, P.O. Box 257, Comptche, California.
22. Wald, H., Shojania, K.G., Root causes analysis, in *Making Healthcare Safer: A Critical Analysis of Patient Safety Practices*, ed. A.J. Markowitz, Report No. 43,

Agency for Healthcare Research and Quality, U.S. Department of Health and Human Services, Rockville, MD, 2001, Chapter 5, pp. 1–7.
23. Dhillon, B.S., Singh, C., *Engineering Reliability: New Techniques and Applications*, John Wiley and Sons, New York, 1981.
24. Dhillon, B.S., *Design Reliability: Fundamentals and Applications*, CRC Press, Boca Raton, Florida, 1999.

# 8

# *Robot Maintenance and Areas of Robotics Applications in Maintenance and Repair*

## 8.1 Introduction

Robots are quite complex machines that can sense information from their surrounding environment, autonomously make effective decisions, and manipulate the surrounding environment with effectors. Just as in the case of any other complex engineering systems, robots also require preventive and corrective maintenance. Therefore, it is very important that users of robots devise an effective maintenance program as, otherwise, the unscheduled downtime of robots may increase. Furthermore, a poor maintenance program may result in safety problems as well.

Needless to say, both robot users and manufacturers play an important role in the maintenance aspects of robots as careful consideration to maintenance is given not only during the operational phase of robots but during the design phase too.

Today, robots are being used in a variety of applications in maintenance and repair quite successfully, among these the nuclear industry, railways, in maintenance of power lines, highways, and in aircraft servicing. Needless to say, this chapter presents various different aspects of robot maintenance and maintenance and repair in different areas of robotics applications.

## 8.2 Robot Maintenance-Related Needs and Maintenance Types

Generally, the maintenance needs of a robot are determined by its specific application and type. The robot powering system is probably the most important factor that clearly affects its maintenance. Most of the robots used in industry sectors may be classified under the following two categories [1]:

- *Hydraulic with electrical controls.* These are robots whose working parts are driven hydraulically and, generally, are controlled by electrical components or parts.

- *Electrical.* These are robots that are driven and controlled by electrically powered parts.

Nonetheless, it is to be noted that irrespective of the robot type, the mechanical parts of a robot require proper attention. Normally, as per [2], the following maintenance-related tasks are relevant to robots:

- Attending to seals and replacing protective accessories
- Inspecting all involved parts regularly for wear, particularly wrists
- Cleaning to eradicate corrosive agents
- Lubricating

Specifically, in the case of hydraulic robots, the items that need proper attention include bearings, servo valves, filters, hydraulic oil, and high-pressure hoses.

Three basic types of maintenance for robots used in industrial applications are corrective maintenance, preventive maintenance, and predictive maintenance. Corrective maintenance is concerned with repairing the robot system to return its operational state whenever it breaks down. Preventive maintenance is basically concerned with servicing robot system parts regularly. Finally, predictive maintenance is concerned with predicting when a failure might occur and to alert personnel involved with maintenance. Predictive maintenance is becoming increasingly effective as modern robot systems are equipped with highly sophisticated electronic parts and sensors.

## 8.3 Commonly Used Tools to Maintain a Robot and Measuring Instruments and Tooling for Periodic Robot Inspections

Various types of tools are used to maintain a robot system. They range from diagnostic codes displayed on the robot control panel to wrenches and specific maintenance manuals. Although the type of maintenance tools required will be peculiar to the specific robot system involved, some of the commonly used tools to maintain a robot system are as follows [1]:

- Torque wrenches
- Circuit-card pullers
- Alignment fixtures

- Seal compressors
- Accumulator charging adaptors

Usually, a variety of measuring instruments and tooling are required to perform periodic inspections on robots. Some examples of measuring instruments are slide calipers, the direct current (DC) voltmeter, the alternating current (AC) voltmeter, and the oscilloscope [3]. Tooling includes items such as the following [3]:

- Large, medium, and small cross-headed screwdrivers
- Grease gun
- Medium and small monkey wrenches
- Double-head wrench
- Large, medium, and small screwdrivers
- Nippers
- Radio pliers
- Pincers
- C-shaped snap-ring pliers
- Hexagon wrench kit

## 8.4 Robot Diagnosis and Monitoring Approaches

In the industrial sectors, many different methods are used for monitoring the robotics equipment status, diagnosing their conditions, and then initiating appropriate maintenance-related activities, such as overhauls, repair, and replacement, when necessary [4]. In this regard, condition-based maintenance policies typically schedule inspections by the involved maintenance or operating personnel on a regular basis. Control charts and other approaches used in the area of quality control are also utilized for determining the degree of deterioration over a period of time in production and for consequent equipment maintenance activities such as replacing parts, recalibration, nonroutine inspections, and adjustments.

Other commonly used diagnostic approaches are vibration analysis, behavior monitoring as well as monitoring of structural integrity, wear debris (i.e., contaminants) hydraulic fluids and lubricants, and so on. Finally, it is to be noted that industrial robots generally contain internal diagnostic functions, which are considered quite useful during troubleshooting and highlight the need to perform maintenance on the machine/system [4].

## 8.5 Useful Guidelines to Safeguard Robot Maintenance Personnel and Safeguarding Methods for Use during the Robot Maintenance Process

Although there are many guidelines to safeguard robot maintenance personnel, the guidelines considered most useful are as follows [5]:

- Ensure that the entire robot system is turned off during the maintenance and repair process, and all the power sources and the releasing of potentially dangerous stored energy are properly locked out/tagged.
- Ensure that all involved maintenance personnel are properly trained in regard to procedures essential to perform all the required maintenance-related tasks safely and effectively.
- Ensure that in the event of a lockout/tagout procedure, equally effective alternative safeguarding methods are employed.
- Ensure that all involved maintenance personnel are properly safeguarded from unintended or unexpected motion.

In situations when power cannot be turned off during the robot system maintenance process, the safeguarding methods shown in Figure 8.1 can be used [6].

The robot arm predetermined position method is concerned with placing the arm of the robot in a predetermined position so that necessary maintenance can be performed without exposing involved personnel to trapping points. The emergency stop method is concerned with making the emergency stop readily accessible, as well as making the restarting of the robot impossible until the time that the emergency stop device is reset manually.

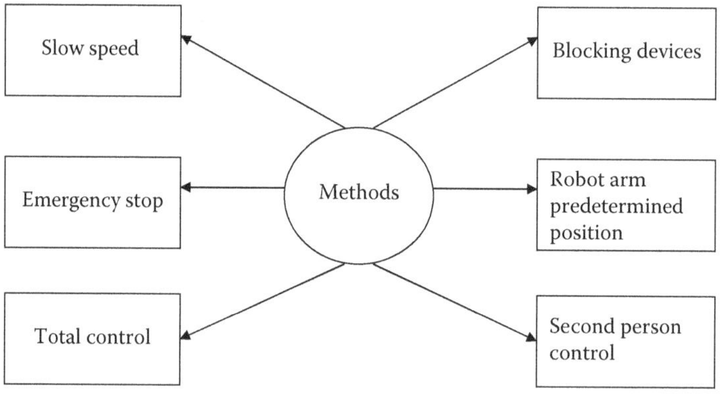


**FIGURE 8.1**
Safeguarding methods for use during the robot system maintenance process.

The total control method is basically concerned with placing the complete control of the robot system in the hands of maintenance personnel. Thus, in this method, the requirements such as those presented below must be considered:

- All robot system-associated emergency stop devices are functional.
- All remote interlocks that can cause robot movement in automatic mode are subject to control override.
- In situations where the other equipment movement in a robot system is hazardous to humans in the restricted work enclosure, the control of such movement is to be placed in the hands of all those individuals.
- The robot system is reset in automatic mode only after the departure of all individuals from the work enclosure.

The slow-speed method is concerned with reducing the speed of robot to a slow-speed level for maintenance operations. The second person control method calls for the presence of a second person (i.e., a person who is knowledgeable about potential hazards associated with a robot system and is also capable of reacting fast to protect others in the moment of need) at the robot system controls.

Finally, the blocking devices method is concerned with using devices such as blocks and pins during the maintenance process for thwarting potentially hazardous moments of the robot system.

## 8.6 Models for Performing Robot Maintenance Analysis

This section presents four mathematical models considered useful, directly or indirectly, to perform robot maintenance analysis.

### 8.6.1 Model I

This mathematical model is concerned with estimating the optimum number of inspections per robot facility per unit time. This type of information is important to decision makers because inspections are often disruptive; however, such inspections generally help to reduce the robot downtime because they result in fewer breakdowns. In this model, the total downtime of the robot is minimized to obtain the optimum number of inspections.

The total robot downtime per unit time is expressed by [7,8]

$$T_{rd} = kDT_i + \frac{(cDT_b)}{k} \tag{8.1}$$

where

$T_{rd}$ is the total robot downtime per unit time.
$DT_i$ is the downtime per inspection for a robot facility.
$c$ is a constant for a specific robot facility.
$k$ is the number of inspections per robot facility per unit time.
$DT_b$ is the downtime per breakdown for a robot facility.

After differentiating Equation 8.1 with respect to $k$ and then equating it to zero, we obtain

$$k^* = \left[\frac{(cDT_b)}{DT_i}\right]^{1/2} \tag{8.2}$$

where

$k^*$ is the optimum number of inspections per robot facility per unit time.

By inserting Equation 8.2 into Equation 8.1, we obtain

$$T_{rd}^* = 2[cDT_b DT_i]^{1/2} \tag{8.3}$$

where

$T_{rd}^*$ is the minimum total robot downtime.

**EXAMPLE 8.1**

Assume that for a certain robot facility, the following data values are given:

$$c = 2$$
$$DT_b = 0.10 \text{ month}$$
$$DT_i = 0.02 \text{ month}$$

Using the above-given data values, calculate the optimum number of monthly robot inspections and the minimum total robot downtime.

By inserting the above-given data values into Equations 8.2 and 8.3, we obtain

$$k^* = \left[\frac{\{2(0.10)\}}{(0.02)}\right]^{1/2}$$
$$= 3.16 \text{ inspections per month}$$

and

$$T_{rd}^* = 2[2(0.10)(0.02)]^{1/2}$$
$$= 0.126 \text{ month}$$

Thus, the optimum number of monthly robot inspections and the minimum total robot downtime are 3.16 and 0.126 month, respectively.

### 8.6.2 Model II

This mathematical model is concerned with maximizing the income of a robot subject to failure and repair. Using the Markov method given in Chapter 4 and [1,9], we write the following expressions for availability and unavailability of a robot:

$$AV_r(t) = \frac{\mu_r}{\lambda_r + \mu_r} + \frac{\lambda_r}{\lambda_r + \mu_r} e^{-(\lambda_r + \mu_r)t} \tag{8.4}$$

$$UAV_r(t) = \frac{\lambda_r}{\lambda_r + \mu_r}[1 - e^{-(\lambda_r + \mu_r)t}] \tag{8.5}$$

where

$UAV_r(t)$ is the robot unavailability at time $t$.
$AV_r(t)$ is the robot availability at time $t$.
$\lambda_r$ is the robot constant failure rate.
$\mu_r$ is the robot constant repair rate.

For very large time $t$, Equations 8.4 and 8.5 reduce to the following equations:

$$AV_r = \lim_{t\to\infty} AV_r(t) = \frac{\mu_r}{\lambda_r + \mu_r} \tag{8.6}$$

$$UAV_r = \lim_{t\to\infty} UAV_r(t) = \frac{\lambda_r}{\lambda_r + \mu_r} \tag{8.7}$$

where

$AV_r$ is the robot steady-state availability.
$UAV_r$ is the robot steady-state unavailability.

Since $\lambda_r = 1/MTTF_r$ and $\mu_r = 1/MTTR_r$, Equations 8.6 and 8.7 become

$$AV_r = \frac{MTTF_r}{MTTF_r + MTTR_r} \tag{8.8}$$

where

$MTTF_r$ is the robot mean time to failure.
$MTTR_r$ is the robot mean time to repair.

and

$$UAV_r = \frac{MTTR_r}{MTTF_r + MTTR_r} \tag{8.9}$$

More specifically, it is to be noted that Equation 8.8 yields the probability that the robot is functional (or a fraction of the time that the robot repair personnel are idle) and Equation 8.9 offers the probability that the robot is under repair (or a fraction of the time that the robot repair personnel are working).

Thus, the monthly cost of the robot maintenance personnel is expressed by

$$C_{rm} = \theta_r \mu_r = \frac{\theta_r}{MTTR_r} \tag{8.10}$$

where

$\theta_r$ is the robot maintenance cost (constant) dependent on the nature of the robot.

The expected monthly income from the robot output is expressed by

$$\begin{aligned} EMI_r &= (MI_{rf})(AV) \\ &= \frac{MI_{rf}(MTTF_r)}{MTTF_r + MTTR_r} \end{aligned} \tag{8.11}$$

where

$MI_{rf}$ is the monthly income from the robot output, if the robot worked full-time.

Thus, the robot's net income is given by

$$\begin{aligned} NI_r &= EMI_r - C_{rm} \\ &= \frac{MI_{rf}(MTTF_r)}{MTTF_r + MTTR_r} - \frac{\theta_r}{MTTR_r} \end{aligned} \tag{8.12}$$

To maximize the robot net income, we differentiate Equation 8.12 with respect to $MTTR_r$, and then set the resulting derivatives equal to zero:

$$\frac{d(NI_r)}{d(MTTR_r)} = \frac{MI_{rf}(MTTF_r)}{(MTTF_r + MTTR_r)^2} + \frac{\theta_r}{(MTTR_r)^2} = 0 \tag{8.13}$$

By rearranging Equation 8.13, we get

$$MTTR^* = \frac{MTTF_r}{[(MI_{rf}(MTTF_r))/\theta_r]^{1/2} - 1} \tag{8.14}$$

where

$MTTR_r^*$ is the optimum mean time needed to repair the robot.

Inserting Equation 8.14 into Equations 8.8, 8.10, and 8.11, respectively, results in

$$AV_r^* = 1 - \left[\frac{\theta_r}{MI_{rf}(MTTF_r)}\right]^{1/2} \tag{8.15}$$

$$C_{rm}^* = \left[\frac{\theta_r MI_{rf}}{MTTF_r}\right]^{1/2} - \frac{\theta_r}{MTTF_r} \tag{8.16}$$

and

$$EMI_r^* = MI_{rf} - \left[\frac{\theta_r MI_{rf}}{MTTF_r}\right]^{1/2} \tag{8.17}$$

where

$AV_r^*$, $C_{rm}^*$, and $EMI_r^*$ are the optimum values of $AV_r$, $C_{rm}$, and $EMI_r$, respectively.

### 8.6.3 Model III

This mathematical model is concerned with a robot that is subjected to a periodical inspection–repair process at a time interval of, say, $T$. The model assumes that at the start of the time interval $T$, the robot is in good condition, but during the interval it could malfunction. Thus, the probability of robotic failure during the time interval $T$ is expressed by [1,10,11]

$$F_r(T) = \int_0^T f(x)dx \tag{8.18}$$

where

$f(x)$ is the probability density function associated with the time between robot malfunctions.

The mean or average availability of the robot over the time interval $T$ is expressed by

$$AV_m = R_r(T) + \frac{X^*}{T} F_r(T) \tag{8.19}$$

where

$R_r(T) = 1 - F_r(T)$ is the probability that the robot will not malfunction during the time interval $T$.

$X^*$ is the time at which the robot malfunction or failure occurs (i.e., within the time interval $T$).

The expected value of $X^*$ is expressed by

$$E(X^*) = \frac{1}{F_r(T)} \int_0^T x\, f(x)dx$$
$$= \frac{TF(T)}{F_r(T)} - \frac{1}{F_r(T)} \int_0^T F(x)dx \tag{8.20}$$

where

$F(x)$ is the cumulative distribution function.

Taking the expected value of Equation 8.19, we obtain the following equation for the robot's expected mean or average availability:

$$E(AV_m) = R_r(T) + \frac{F_r(T)E(X^*)}{T} \tag{8.21}$$

By inserting Equation 8.20 into Equation 8.21, we get

$$E(AV_m) = [1 - F_r(T)] + \frac{F_r(T)}{T}\left[\frac{TF(T)}{F_r(T)} - \frac{1}{F_r(T)} \int_0^T F(x)dx\right]$$
$$= 1 - \frac{1}{T} \int_0^T F(x)dx \tag{8.22}$$

It is to be noted that Equation 8.22 is subject to the assumption that the duration of the inspection–repair process is negligible, which may or may not be realistic. Nonetheless, a more realistic assumption for a regular inspection–repair could be a duration of, say, time $T_r$. In such a situation, the net availability of the robot is expressed by

$$AV_{nr} = \frac{(AV_m)T}{T + T_r} \tag{8.23}$$

Thus, the expected value of Equation 8.23 is

$$E(AV_{nr}) = \left(\frac{T}{Y}\right) E(AV_m) \tag{8.24}$$

where
$Y = T + T_r$

By inserting Equation 8.22 into Equation 8.24, we get

$$\begin{aligned} E(AV_{nr}) &= \left(\frac{T}{Y}\right)\left[1 - \frac{1}{T}\int_0^T F(x)dx\right] \\ &= \frac{1}{Y}\left[T - \int_0^T F(x)dx\right] \end{aligned} \tag{8.25}$$

**EXAMPLE 8.2**

Assume that the mean time to failure of a robot is 400 h and its failure times are exponentially distributed. After every 200 h of service, the robot is scheduled for periodic maintenance.

Calculate the robot's expected average availability over a time interval of 200 h under the following assumptions:

- Each and every inspection and repair restores the robot to a condition that is considered as good as new.
- The robot is repaired only at the scheduled time.
- The duration of time taken by the inspection–repair process is negligible.

Thus, in this case, the robot's failures are expressed by the following probability density function:

$$f(x) = \frac{1}{400}e^{-x/400}, \quad x \geq 0 \tag{8.26}$$

where
$x$ is the time to failure random variable.

With the aid of the definition for cumulative distribution function given in Chapter 2, we obtain the following cumulative distribution function of the robot failure time:

$$\begin{aligned} F(x) &= \frac{1}{400}\int_0^x e^{-x/400}dx \\ &= 1 - e^{-x/400} \end{aligned} \tag{8.27}$$

By inserting Equation 8.27 into Equation 8.22, we get

$$\begin{aligned} E(AV_m) &= 1 - \frac{1}{200}\int_0^{200}\left[1 - e^{-x/400}\right]dx \\ &= 1 - \frac{1}{200}\left[x + 400e^{-x/400}\right]_0^{200} \\ &= 0.7869 \end{aligned}$$

Thus, the expected average availability of the robot is 0.7869.

### 8.6.4 Model IV

This mathematical model is concerned with determining the robot's optimum inspection frequency to minimize the robot downtime per unit time. In this model, the total robot downtime (per unit time) is the function of inspection frequency and is expressed by [12,13]

$$\begin{aligned} RTDT(k) &= RDT_i + RDT_r \\ &= \frac{k}{\alpha} + \frac{\lambda_r(k)}{\mu_r} \end{aligned} \tag{8.28}$$

where

$RDT_i$ is the robot downtime due to per-unit-of-inspection.
$RDT_r$ is the robot downtime due to per-unit-of-time repairs.
$RTDT(k)$ is the robot total downtime per unit of time.
$1/\alpha$ is the mean value of exponentially distributed inspection times.
$\lambda_r(k)$ is the robot failure rate.
$\mu_r$ is the robot repair rate.
$k$ is the inspection frequency.

By differentiating Equation 8.28 with respect to $k$, we obtain

$$\frac{dRTDT(k)}{dk} = \frac{1}{\alpha} + \frac{1}{\mu_r} \cdot \frac{d\lambda_r(k)}{dk} \tag{8.29}$$

By setting Equation 8.29 equal to zero and then rearranging, we get

$$\frac{d\lambda_r(k)}{dk} = -\frac{\mu_r}{\alpha} \tag{8.30}$$

It is to be noted that the value of $k$ will be optimum when the right and left sides of Equation 8.30 are equal. At this very point, the robot total downtime will be minimal.

**EXAMPLE 8.3**

Assume that the failure rate of a robot is expressed by

$$\lambda_r(k) = \lambda e^{-k} \tag{8.31}$$

where
$\lambda$ is the robot failure rate at $k = 0$.

Obtain an expression for the optimal value of $k$ with the aid of Equation 8.30 and then—by setting the value of $1/\mu_r = 0.04$ month, $1/\alpha = 0.01$ month, and $\lambda = 2$ failures per month—calculate the optimal value of the inspection frequency, $k$.

By inserting Equation 8.31 into Equation 8.30, we obtain

$$-\lambda e^{-k} = -\frac{\mu_r}{\alpha} \tag{8.32}$$

By rearranging Equation 8.32, we get

$$k* = \ln\left[\frac{\lambda\alpha}{\mu_r}\right] \tag{8.33}$$

where
$k^*$ is the optimal value of inspection frequency, $k$.

By substituting the given data values into Equation 8.33, we get

$$k* = \ln\left[\frac{2(0.04)}{0.01}\right]$$
$$= 2.08 \text{ inspections per month}$$

Thus, the optimal value of the inspection frequency is 2.08 inspections per month.

## 8.7 Areas of Robotics Applications in Maintenance and Repair

Over the years, robots—whether teleoperated, under supervisory control, or autonomous—have been used in or are being considered for many different applications in the areas of maintenance and repair. Some of the areas of robotics applications in maintenance and repair are as follows [14]:

- Nuclear industry
- Highways

- Railways
- Underwater facilities
- Power line maintenance
- Aircraft servicing

The first five of the above robotics applications are described below and additional information on the remaining one is available in [14].

### 8.7.1 Nuclear Industry

Over decades, teleoperated robots have been well utilized in the area of maintenance in the nuclear industry. There are several features of maintenance that make their applications quite attractive in this arena. Some of these features are as follows:

- Repair and maintenance need high levels of dexterity.
- The low frequency of the operation clearly calls for a general-purpose system that is capable of performing an array of maintenance-related tasks.
- The complexity of the tasks could be quite unpredictable due to the uncertain impact of a failure.

Although humans are likely to be able to accomplish tasks faster than teleoperators, the use of teleoperators can be quite useful to

- Lower health-related risks and improve safety
- Improve availability by allowing the maintenance activity to take place during ongoing operations instead of halting ongoing operations
- Reduce mean time to repair by shortening the response time to failures

The main application areas of teleoperated robots in nuclear maintenance are as follows [14,15]:

- *Maintenance in nuclear reactors.* As nuclear reactors are less likely to be designed for robotic maintenance than industrial nuclear facilities, it often calls for innovative approaches to gain access to trouble areas. In the past, such factors have restricted the application of robots in this arena. However, owing to increasingly stringent limits on worker exposure to ionizing radiation and exposure reductions, increasing attention is being given to use robots for maintenance activities.

Additional information on this is available in [15–21].

- *Operation and maintenance of industrial nuclear facilities and laboratories.* Additional information on this area is available in [14–16,22].
- *Emergency intervention.* Additional information on this area is available in [23–25].
- *Dismantling and decommissioning nuclear facilities.* Additional information on this area is available in [26–30].

### 8.7.2 Highways

Highways have become important in the transportation network around the globe. Over the years, the increasing volume of traffic on the roadways causes deterioration of roadways due to poor maintenance. Robotics solutions in highway maintenance applications are becoming increasingly attractive because of their potential for lowering labor costs, improving the safety of the highway workers, increasing the repair quality, reducing delays in traffic flow, and so on.

Some of the applications in which robotics can be used in this specific area are shown in Figure 8.2 [14–31].

Although, over the years, relatively few implementations have been attempted in the area of robotics in highway maintenance and repair, many

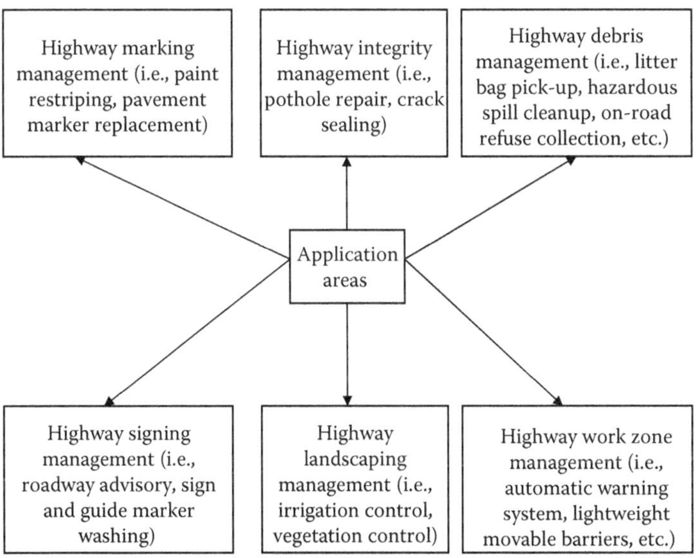


**FIGURE 8.2**
Robotic application areas in highway maintenance.

successful prototypes have been developed [32]. Some examples of robotics developments for the highway maintenance areas are as follows:

- Automated pavement distress data-collection vehicles [14,32]
- An automated cone dispenser that can effectively dispense and remove upto two rows of warning cones, for a total of about 240 cones [14,32]
- A rapid runway repair telerobitic system to repair craters on runways [14,33]
- A mobile lane separator that can place and remove concrete road marker blocks at speeds of upto 30 km/h [14]

### 8.7.3 Railways

Over the years, the railroad sector has recognized the economic advantages of automation, resulting in the development of a number of robotics solutions for the maintenance and repair in railways. The most common locations of robots are the railway maintenance workshops, which undertake activities such as cleaning, painting, and grinding [34]. Two examples of robotics developments in railway maintenance are as follows:

- The development of an automated system for cleaning the undersides of subway cars; the system involves the application of three industrial painting robots installed on either side and under the subway car being cleaned [35]. In this case, an operator starts the cleaning operations and then the complete system is controlled by a master computer that supervises the three individual robot controllers.
- The development of the RMS-2 rail grinding system that has automated capabilities for sensing the existing condition of the rails' surface [36]. The system can grind rails at speeds upto six miles per hour and contains 120 stones for grinding, spread along the underside of five of the 12 railway cars that make up the total system.

### 8.7.4 Underwater Facilities

In the offshore oil excavation sector, teleoperated robots are often used for maintaining facilities beneath the ocean surface. Their specific applications include repairing items such as pipelines, platforms, communications cables, and well heads [14].

Teleoperated robots are also used to inspect and repair water conveyance tunnels, inspect and clean steam generators, carry out underwater construction, clean marine growth from power plant cooling systems, and so on

[37–39]. As per [40], there are around 180 different commercially available, remotely controlled systems and around 63 companies are involved in building remotely operated manipulators or vehicles for subsea work. Needless to say, there is a wide range of remotely operated systems and additional information on the topic is available in [14,37–39].

### 8.7.5 Power Line Maintenance

Human operators perform maintenance-related tasks on power lines such as replacing ceramic insulators that support conductor wire as well as opening and enclosing the circuit between poles on overhead transmission lines. In order to replace humans in performing tasks such as these, many electric power companies have, since the 1980s, been investigating the use of robotic systems for live-line power line maintenance. Some examples of the robotics developments in power line maintenance are as follows:

- Teleoperator for operations, maintenance, and construction using advanced technology (TOMCAT) [41]
- A two-manipulator telerobotic system for live-line maintenance [42]
- ROBTET teleoperated system for live-line maintenance [43]
- Robot for automatic washing and brushing of polluted electric insulators [44]

## 8.8 Problems

1. Discuss robot maintenance-related needs and maintenance types.
2. What are the commonly used tools to maintain a robot?
3. Discuss robot diagnosis and monitoring approaches.
4. What are the most useful guidelines to safeguard robot maintenance personnel?
5. What are the safeguarding methods that can be used during the robot maintenance process?
6. Prove Equation 8.15 by using Equation 8.12.
7. Prove Equation 8.30 with the aid of Equation 8.28.
8. Discuss the following two areas of robotics applications in maintenance and repair:
   a. Nuclear industry
   b. Railways

9. Discuss the following three areas of robotics applications in maintenance and repair:
   a. Highways
   b. Power line maintenance
   c. Underwater facilities
10. List at least 10 tooling items that can be used for periodic robot inspections.

## References


1. Dhillon, B.S., *Robot Reliability and Safety*, Springer-Verlag, New York, 1991.
2. Lester, W.A., Lannon, R.P., Bellandi, R., Robot users need to have a program for maintenance, industrial engineering, Part I, *Industrial Engineering*, Vol. 17, 1985, pp. 28–32.
3. Japanese Ministry of Labor, Japanese Industrial Safety and Health Association, ed., *An Interpretation of the Technical Guidance on Safety Standards in the Use, Etc., of Industrial Robots*, Tokyo, 1985.
4. Clark, D.R., Lehto, M.R., Reliability, maintenance, and safety of robots, in *Handbook of Industrial Robotics*, ed. S.Y. Nof, John Wiley and Sons, New York, 1999, pp. 717–753.
5. Robotics Industries Association, *American National Standard for Industrial Robots and Robot Systems: Safety Requirements, Document No. ANSI/RIA R15.06, 1986*, 900 Victors Way, P.O. Box 3724, Ann Arbor, Michigan 48016, 1986.
6. Lodge, J.E., How to protect robot maintenance workers, *National Safety News*, 1984, pp. 48–51.
7. Wild, R., *Essentials of Production and Operations Management*, Holt, Rinehart, and Winston, London, 1985.
8. Dhillon, B.S., *Mechanical Reliability: Theory, Models, and Applications*, American Institute of Aeronautics and Astronautics, Washington, DC, 1988.
9. Dhillon, B.S., *Design Reliability: Fundamentals and Applications*, CRC Press, Boca Raton, Florida, 1999.
10. Pages, A., Gondran, M., *System Reliability*, Springer-Verlag, New York, 1986.
11. Morse, P.M., *Queues, Inventories, and Maintenance*, John Wiley and Sons, New York, 1958.
12. Jardine, A.K.S., *Maintenance, Replacement, and Reliability*, Pitman Publishing, London, 1973.
13. Dhillon, B.S., *Engineering Maintenance: A Modern Approach*, CRC Press, Boca Raton, Florida, 2002.
14. Parker, L.E., Draper, J.V., Maintenance and repair, in *Handbook of Industrial Robotics*, ed. S.Y. Nof, John Wiley and Sons, New York, 1999, pp. 1023–1036.
15. Vertut, J., Coiffet, P., *Teleoperation and Robotics: Applications and Technology*, Prentice-Hall, Englewood Cliffs, 1985.

16. Chesser, J.B., *BRET Rack Remote Maintenance Demonstration Test, Report No. ORNL/TM-10875*, Oak Ridge National Laboratory, Oak Ridge, 1988.
17. Lovett, J.T., Development robotic maintenance systems for nuclear power plants, *Nuclear Plant Journal*, Vol. 9, 1991, pp. 87–90.
18. Macdonald, D. et al., Remote replacement of TF and PF coils for the compact ignition Tokamak, *Proceedings of the 4th ANS Topical Meeting on Robotics and Remote Systems*, 1991, pp. 131–140.
19. Tsukune, H. et al., Research and development of advanced robots for nuclear power plants, *Bulletin of the Electrochemical Laboratory*, Vol. 58, No. 4, 1994, pp. 51–65.
20. Glass, S.W. et al., Modular robotic applications in nuclear power plant maintenance, *Proceedings of the 58th American Power Conference*, 1996, pp. 421–426.
21. Ali, M., Puffer, R., Roman, H., Evaluation of a multifingered robot hand for nuclear power plant operations and maintenance tasks, *Proceedings of the 5th World Conference on Robotics Research*, 1994, pp. 217–227.
22. Horne, R.A. et al., Extended tele-robotic activities at CERN, *Proceedings of the 4th ANS Topical Meeting on Robotics and Remote Systems*, 1991, pp. 525–534.
23. Merchant, D.J., Tarpinian, J.E., Post-accident recovery operations at TMI-2, *Proceedings of the Workshop on Requirements of Mobile Teleoperators for Radiological Emergency Response and Recovery*, 1985, pp. 1–9.
24. Bengal, P.R., The TMI-2 remote technology program, *Proceedings of the Workshop on Requirements of Mobile Teleoperators for Radiological Emergency Response and Recovery*, 1985, pp. 49–60.
25. Chester, C.V., Characterization of radiological emergencies, *Proceedings of the Workshop on Requirements of Mobile Teleoperators for Radiological Emergency Response and Recovery*, 1985, pp. 37–48.
26. Noakes, M.W., Haley, D.C., Willis, W.D., The selective equipment removal system dual arm work module, *Proceedings of the 7th Topical Meeting on Robotics and Remote Systems*, 1997, pp. 478–483.
27. Randolph, J.D. et al., Development of waste dislodging and retrieval system for use in the Oak Ridge National Laboratory Gunnite Tanks, *Proceedings of the 7th Topical Meeting on Robotics and Remote Systems*, 1997, pp. 894–906.
28. Kiebel, G.R., Carteret, B.A., Niebuhr, D.P., Light duty utility arm deployment in Hanford Tank T-106, *Proceedings of the 7th Topical Meeting on Robotics and Remote Systems*, 1997, pp. 921–930.
29. Draper, J.V., Function analysis for the single-shell tank waste retrieval manipulator system, Report No. ORNL/TM-12417, Oak Ridge National Laboratory, Oak Ridge, 1993.
30. Draper, J.V., Task analysis for the single-shell tank waste retrieval manipulator system, Report No. ORNL/TM-12432, Oak Ridge National Laboratory, Oak Ridge, 1993.
31. Ravani, B., West, T., Applications of robotics and automation in highway maintenance operations, *Proceedings of the 2nd International Conference on Applications of Advanced Technologies in Transportation Engineering*, 1991, pp. 61–65.
32. Zhou, T., West, T., Assessment of the state-of-the-art of robotics applications in highway construction and maintenance, *Proceedings of the 2nd International Conference on Applications of Advanced Technologies in Transportation Engineering*, 1991, pp. 56–60.

33. Nease, A., Development of Rapid Runway Repair (RRR) telerobotic construction equipment, *Proceedings of the IEEE National Telesystems Conference*, 1991, pp. 321–322.
34. Martland, C., Analysis of the potential impacts of automation and robotics on locomotive rebuilding, *IEEE Transactions on Engineering Management*, Vol. 34, No. 2, 1987, pp. 92–100.
35. Wiercienski, W., Leek, A., Feasibility of robotic cleaning of undersides of Toronto's subway cars, *Journal of Transportation Engineering*, Vol. 116, No. 3, 1990, pp. 272–279.
36. Toward the "Smart" rail maintenance system, *Railway Track & Structures*, Vol. 82, No. 11, 1986, pp. 21–24.
37. Edahiro, K., Development of "Underwater Robot" cleaner for live growth in power station, in *Teleoperated Robotics in Hostile Environments*, eds. H.L. Martin and D.P. Kuban, Robotics International Society of Manufacturing Engineers, Dearborn, Michigan, 1985, pp. 108–118.
38. Travato, S.A., Ruggieri, S.K., Design, development and field testing of CECIL: A steam generator secondary side maintenance robot, *Proceedings of the 4th ANS Topical Meeting on Robotics and Remote Systems*, 1991, pp. 121–130.
39. Yemington, C., Telerobotics in the offshore oil industry, *Proceedings of the 4th ANS Topical Meeting on Robotics and Remote Systems*, 1991, pp. 441–450.
40. Gallimore, D., Madsen, A., *Remotely Operated Vehicles of the World*, Herfordshire: Oilfield Publications, Ledbury, U.K., 1994.
41. Dunlap, J., Robotic maintenance of overhead transmission lines, *IEEE Transactions on Power Delivery*, Vol. 1, No. 3, 1986, pp. 280–284.
42. Yano, K. et al., Development of the semi-automatic hot-line work robot system "Phase II," *Proceedings of the 7th International Conference on Transportation and Distribution Construction and Live Line Maintenance*, 1995, pp. 213–218.
43. Aracil, R. et al., ROBTET: A new teleoperated system for live-line maintenance, *Proceedings of the 7th International Conference on Transmission and Distribution Construction and Live-Line Maintenance*, 1995, pp. 205–211.
44. Yi, H., Jiansheng, S., The research of the automatic washing-brushing robot of 500 kv DC insulator string, *Proceedings of the 6th International Conference on Transmission and Distribution Construction and Live Line Maintenance*, 1993, pp. 411–424.

# 9

# *Human Factors and Safety Considerations in Robotics Workplaces*

## 9.1 Introduction

Human factors and safety are important in the design of robotic workplaces, as they are intimately related. More clearly, many safety problems have solutions in human factors.

Generally, in regard to robotics, human factors are basically concerned with the way robotic equipment interface systems and the environment should be designed so that they are effectively compatible with the humans who use it. In industrial robotics, in addition to ambient conditions, basic concern is associated with the design of items such as controller panels, teach pendants, and computer terminals [1].

Needless to say, one of the major motives for investing in industrial robotics is to relieve human operators from tasks that are hazardous or difficult. Over the years, significant progress has been made in the area of human factors and safety considerations in robotics workplaces. This chapter presents various important aspects of human factors and safety considerations in robotics workplaces.

## 9.2 Human Factors-Related Issues during the Robotic Systems' Factory Integration Process

During the factory integration of robotic systems, there are many issues related to human factors that must be taken into consideration with care. The primary eight of these issues are the following [2,3]:

- Safety
- Maintainability
- Worker–machine interface

- Selection and training
- Work environment
- Management
- Job design
- Communication among workers and between workers and management

Each of the above issues is described below, separately.

### 9.2.1 Safety

The safety of workers and other individuals is a very serious issue and must be considered with care during the factory integration process of robotic systems. Three aspects to robot-related safety are as follows [2,3]:

- Providing appropriate protection to robot operators, maintenance personnel, and programmers
- Providing appropriate protection to the robots themselves
- Providing safety to curious outsiders

Furthermore, there are several issues that require careful attention. These include designing an effective production layout with regard to safety, reducing the effect of environmental hazards, and labeling potential hazards.

### 9.2.2 Maintainability

Robot maintainability is an important issue, as diagnostic-related activities can account for around 70% of maintenance time. Thus, some of the useful human factors-related guidelines/notes for the maintenance of robotic systems are presented below [2,3]:

- Ensure that robotic system maintenance is performed by personnel who work in the same environment as the robots.
- Maintenance personnel require appropriate multicraft capability as well as skills in robot system operation and programming knowledge for carrying out diagnoses and tests.
- The response time of maintenance personnel to a robotic workstation could be affected by a large number of factors, including craft coordination, the availability of qualified personnel, and the inclusion of robots on a priority list for maintenance.
- Appropriate maintenance manuals are required to carry out robotic maintenance effectively.

- Various specialties are needed for robotic maintenance (e.g., oiling, mechanical, electronic, and plumbing).
- Ensure that well-qualified maintenance personnel are provided in all operating shifts to have effective production results.

### 9.2.3 Worker–Machine Interface

The main objective of the worker–machine interface issue is to design equipment to match capabilities and shortcomings of concerned workers effectively. In regard to robotics, as robot systems are maintained, operated, or programmed by humans, they are man–machine systems.

Two useful worker–machine interface guidelines concerning the factory integration of robotics are as follows:

- During the interface design phase, consider with care items such as the tasks of the robot maintenance personnel, operators, and programmers, and the location of the controls.
- For each and every robot application and man–machine interactive task, determine the most suitable speed range for the robot. In addition, aim to make the robot speed (i.e., pace rate) be operator-paced.

### 9.2.4 Selection and Training

The issue of selection and training is concerned with the robot maintenance personnel, operators, and programmers. It involves selecting and locating appropriate workers, placing the workers appropriately for the relevant job, and training or retraining those workers to have the necessary skill, knowledge, and ability.

Finally, an effective training program requires careful consideration as to the training objective, the training method, the evaluation approaches, and the needs of the job or task.

### 9.2.5 Work Environment

The work environment incorporates both the physical and psychological environments under which a worker concerned with robotic systems performs his/her task; this includes issues of the potential for experiencing discomfort in the given environment. The physical environment includes items such as glare, noise, low ambient illumination, and an incorrect work position.

Alongside, the psychological environment includes items such as improperly devised rest periods, disagreeable coworkers, and monotonous and hectic tasks.

### 9.2.6 Management

Issues of management include elements such as the following:

- The effect of management decisions on the morale and motivation of workers concerned with robotics systems
- Employee relations
- Organizational design

### 9.2.7 Job Design

The issue of job design is a quite challenging human factor issue. The objective is to design jobs/tasks such that the individual differences are clearly recognized, in addition to the maximization of individual productivity and satisfaction.

Here, it is to be noted that the performance of task analysis is very important during designing jobs.

### 9.2.8 Communication among Workers and between Workers and Management

The issue of communications is concerned with establishing effective formal and informal communication channels among various groups of people, for example, between robot maintenance personnel and operators, between robot operators on different shifts, and between robot programmers and operators.

It is to be noted with care that effective communication among workers and between workers and management personnel is very important for the successful factory integration of robotic systems.

## 9.3 Common Robot and Robot-Related Human Tasks

Robots are used to perform various types of tasks. The common ones are shown in Figure 9.1 [4].

Similarly, the common robot-related human tasks are as follows [4]:

- *Maintenance.* This human task is concerned with maintenance of robots; it is complex and generally requires knowledge of mechanics, electronics, hydraulics, and computers.
- *Programming.* This human task is concerned with programming of robots, commonly involving the following:
  - By using a teach pendant
  - By manually leading the robot arm
  - Through a computer terminal

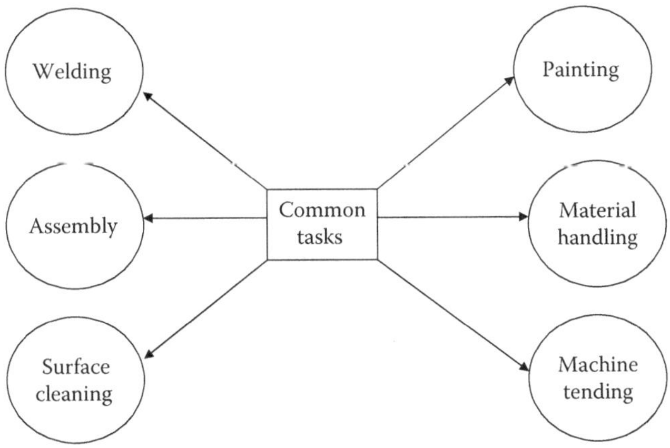


**FIGURE 9.1**
Common robot tasks.

- *Monitoring/supervising.* This robot-related human task involves the detection of events that require operator attention. In this case, the operator has two options in monitoring the situation: directly monitoring by observing the robot or indirectly monitoring by observing displays.
- *Intervening/helping.* This human task is concerned with intervening/helping, as robots often require help to carry out complex tasks. For example, in robotic assembly, gravity-feed devices have to be cleared and items that the robot drops have to be picked up from the floor; thus, in this case, the robot operator is helping/serving the robot.
- *Leftover tasks.* These human tasks are those that the robot has difficulty in performing. One example of such tasks is visual inspection and pattern recognition.
- *Replace robot during failure.* This human task is concerned with replacing the robot during failure in performance as sometimes it becomes essential for operators to replace robots.

## 9.4 Rules of Robotics in Regard to Humans and Advantages and Disadvantages of Robotization with Respect to Human Factors

Over the years, many rules of robotics with regard to humans have been established. Seven of these rules are as follows [3,5]:

- Robots must be easily operable by all involved humans.

- Robots must only replace humans on jobs/tasks that are considered to be hazardous or which the human beings are clearly unwilling to undertake.
- Robots must not psychologically and physically oppress human beings.
- Robots must be used and produced with the objective of contributing to the welfare and development of human beings.
- Whenever robots replace humans in certain jobs/tasks, obtain the prior approval of those who will be affected.
- All robots must be under the command of human beings, so that other human beings are not harmed and only robots themselves are damaged.
- At the accomplishment of assigned tasks or jobs, robots must leave the area to avoid interference with humans and other robots in the area.

There are many advantages and disadvantages of robotization, with respect to human factors. Some of the advantages are as follows [3,6]:

- To free workers from having to carry out tasks involving a heavy physical load
- To free workers from carrying out difficult and monotonous jobs/tasks
- To free workers from having to carry out jobs/tasks in the presence of high noise, vibrations, and so on.
- To free workers from carrying out jobs/tasks at high temperatures
- To free workers from having to carry out microscopic tasks/jobs that need acute eyesight
- To free workers from having to carry out tasks/jobs in the environment of harmful gases
- To supplement the capabilities of handicapped and old persons
- To free workers from carrying out jobs/tasks exposed to nuclear radiation

Similarly, some of the disadvantages of robotization with respect to human factors are as follows [3,6]:

- Human safety-related problems
- Difficult to operate because of significant variations of control panel layouts from one root model to another

- Incompatibility in terms of motion with humans— for example, robots move at acute angles or linearly and abruptly stop but humans do not
- Incompatibility with humans in regard to work pace

## 9.5 Humans at Risk from Robots and Risk-Reducing Measures to Prevent Robot-Related Human Accidents

Experiences over the years clearly indicate that there are several people who could be at risk from robots used in assembly-oriented work. These people are categorized under the following four groups [3,7]:

- *Robot programmers.* These people are subject to various types of injuries because they come into direct physical contact with robots. Although, during the programming process, robots move just a small fraction of their normal operating speeds, failures do occur from time to time and programmers may get injured.
- *Maintenance personnel.* These people are in direct physical contact with robots because of the nature of their work. Thus, they could be subject to hand or face injuries, electrocution, and so on.
- *People outside the identified danger zone.* These individuals are also at risk to a certain degree because, under certain circumstances, the robot gripper may fling out parts/components that may, in turn, strike them.
- *Casual observers.* These individuals are also at risk from injury by robots because of their ignorance. For example, they may assume that the robot is in a stationary state when, in fact, it is not.

Some of the risk-reducing measures to prevent robot-related human accidents are as follows [4,8]:

- *Measure I*: Paint the robot with a special color for making its position sharply visible
- *Measure II*: Equip the robot with a touch-sensitive skin so that it automatically shuts down in the event of a physical contact with a human
- *Measure III*: Locate the robot so that all the pinch-points are totally eliminated
- *Measure IV*: Design the robot to lower or eliminate the risk of cuts and pressure contusions

## 9.6 Useful Guidelines to Safeguard Robot Teachers and Operators

The safety of robot teachers is an important issue and it requires a careful consideration. Some of the useful guidelines for safeguarding robot teachers are as follows [3,5]:

- Each robot system user must ensure that all involved robot teachers are properly trained in regard to the recommended "teach" procedures, the control program, and the specific installation
- Require robot teachers to vacate the restricted robot work zone prior to switching to automatic mode
- Function test the teach control of the pendant and repair any damage or failure prior to commencement of the teaching operation
- At the time of choosing the teach mode, fully satisfy conditions such as cutting of the robot from responding to any signals that may cause motion, putting complete control of any other equipment in the prohibited work area in the hands of the robot teacher(s) if their movements would present a potential hazard, and putting the complete robot system under the control of the teacher(s)
- Require robot teachers to properly inspect the robot visually, including its restricted work area, before starting teaching (the robot), in order to ensure that hazard-causing conditions do not exist
- Require robot teachers to ensure that all safeguards are properly in place and functional, as specified, in the teach mode before entry into the prohibited robot work area

Some of the useful guidelines for safeguarding robot operators are as follows [3,5]:

- Provide appropriate safeguards to prevent the entry of robot operators into the prohibited work area while the robot is in motion
- Each robot system user must establish appropriate safeguards for each and every operation associated with the robot system
- Provide appropriate training to robot system operators in the proper operation of the robot system's control actuators
- Provide appropriate training to robot system operators so that they can quickly recognize known hazards associated with each and every task involving the robot system

- Provide appropriate training to robot system operators so that they can respond properly and quickly to recognized hazardous conditions

## 9.7 Approaches for Limiting Robot Movements

There are many approaches that can be used for limitation of robot movements. Six of these approaches are shown in Figure 9.2 [4].

Motion-limiting devices are used to restrict movements of robots by using limiting switches and fixed stops located close to the robot joints. Robot-locking devices such as chains and pins are used to keep the robot in one location in situations when the robot is inactive or has been temporarily halted, due to, say, maintenance work.

Brakes are applied to the joints of the robot, which may stop the movement of robot at any location. An important criterion to evaluate the brakes' effectiveness is the stopping distance. The "watchdog safety computer" is a stand-alone microcomputer system used for monitoring movements of robots in their workstations; its purpose is to detect operations that are outside the normal conditions' range as well as to stop robots before collision or damage.

Manual stops include emergency stop buttons and advance switches; speed limitation for teaching and maintenance approach is considered self-explanatory.

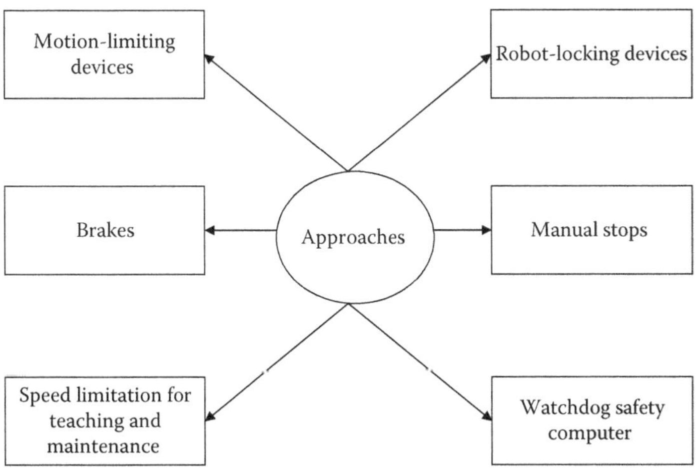


**FIGURE 9.2**
Approaches for limiting robot movement.

## 9.8 Methods for Analysis of Safety and Human Error in Robotics Workplaces

There are many methods developed over the years that can be used to analyze safety and/or human error in robotics workplaces [4,9]. A number of these methods are described below, separately.

### 9.8.1 Deviation Analysis

This method is concerned with analyzing accidents that result from deviations or disturbances in the planned or normal production process. These deviations are applicable to organizational, technical, and individual functions. The method is composed of the following three steps [4,10]:

- *Step 1*: This step is concerned with summarizing the system functions and different activities involved, and then dividing the activities into blocks of suitable size. Furthermore, for each and every block, the normal activities and possible deviations from normal activities are listed.
- *Step 2*: This step is concerned with judging the consequences of the deviations, such as dangerous situations or conditions.
- *Step 3*: This step is concerned with determining the systems improvements that would lower the severity levels of problematic consequences.

### 9.8.2 Energy Barrier Analysis

The basis for this method is that accidents occur due to release or transfer of energy that becomes uncontrollable. The method is composed of the following four steps [4,11]:

- *Step 1*: Highlight all possible sources of potential unwanted energy (PUE) in the robot-associated environment
- *Step 2*: Highlight effective mechanisms for lowering and controlling PUE levels from the source
- *Step 3*: Highlight effective mechanisms to block the transfer of PUE
- *Step 4*: Highlight effective mechanisms to warn the operator of the existence of PUE

It is to be noted that PUE is generally generated from environmental products that the robot is operating with (i.e., chemical, thermal, etc.), or from the robot (hydraulic or electrical power) itself. As per [11], the most practical barrier solution approach may be to block the transfer of PUE to the human operator. This may imply making use of devices such as motion sensors, vision sensors, and floor mats, all of which must be sensitive to human proximity.

### 9.8.3 Task Analysis

This method is concerned with breaking down the main task into subtasks and then into task elements. These may then be analyzed in regard to information processing, information input and output. In situations where there is cooperation between humans and robots, it may be quite appropriate to carry out a dual task analysis of the human-related and robot-related actions in combination [4]. A mathematical approach for modeling the reliability of a system, where actions are executed intermittently by the human and the robot, is available in [12].

Additional information on this method is available in [4,13].

### 9.8.4 Fault Tree Analysis

This is a very useful method and, over the years, has been used to analyze the safety and reliability of human–machine systems. In 1983, National Institute for Occupational Safety and Health (NIOSH) proposed this method to analyze maintenance aspects of industrial machines [14].

As per past experiences, since maintenance is the primary activity that has been implicated in the occurrence of robot accidents, fault tree analysis is considered highly relevant [4]. The method is described in Chapter 4 and in extensive detail in [15–16]. The application of the fault tree method to robot accidents in workplaces is demonstrated through the simple example presented below [3].

**EXAMPLE 9.1**

Assume that in a robotics workplace a robot accident involving a human can only occur due to two reasons: human in the robotics work zone and a sudden movement of the robot. The presence of the human in the robotics work zone could be either due to intentional entry or unintentional entry. Similarly, the sudden movement of the robot(s) can happen due to two reasons: power supply on and unexpected start signal.

Develop a fault tree for the top event: robot accident involving a human and then calculate the probability of occurrence of the top event if the probabilities of occurrence of events (i.e., reasons) are, respectively, 0.05, 0.15, 0.2, and 0.1, for the power supply being on, an unexpected start signal, intentional entry, and unintentional entry.

By using the fault tree symbols given in Chapter 4, a fault tree for the example is shown in Figure 9.3. The single capital letters in the figure denote corresponding fault events (i.e., *T*: robot accident involving a human, *A*: sudden movement of robot, *B*: human in the robotics work zone, *C*: power supply on, *D*: unexpected start signal, *E*: intentional entry, *F*: unintentional entry).

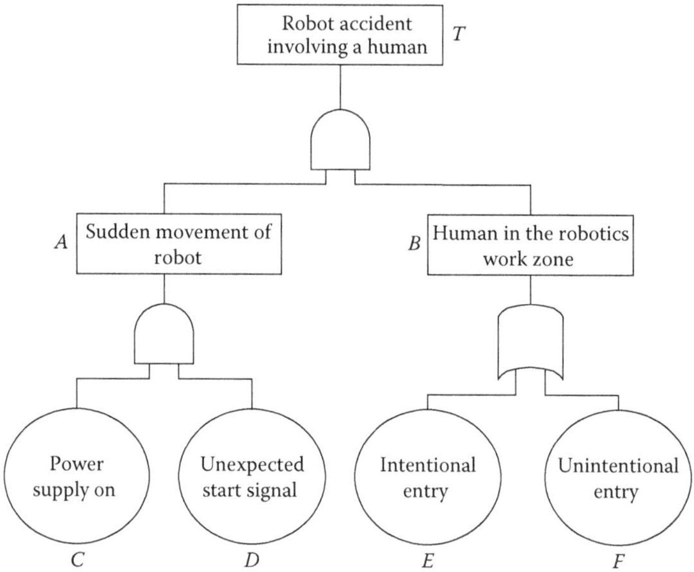


**FIGURE 9.3**
A fault tree for Example 9.1.

With the aid of Chapter 4, material concerning fault tree analysis and the given data values, the probability of occurrence of event $A$ is

$$\begin{aligned} P(A) &= P(C)P(D) \\ &= (0.05)(0.15) \\ &= 0.0075 \end{aligned} \tag{9.1}$$

where
$P(C)$ is the probability of occurrence of event $C$.
$P(D)$ is the probability of occurrence of event $D$.

Similarly, with the aid of Chapter 4, material concerning fault tree analysis and the given data values, the probability of occurrence of event $B$ is

$$\begin{aligned} P(B) &= P(E) + P(F) - P(E)P(F) \\ &= 0.2 + 0.1 - (0.2)(0.1) \\ &= 0.28 \end{aligned} \tag{9.2}$$

where
$P(E)$ is the probability of occurrence of event $E$.
$P(F)$ is the probability of occurrence of event $F$.

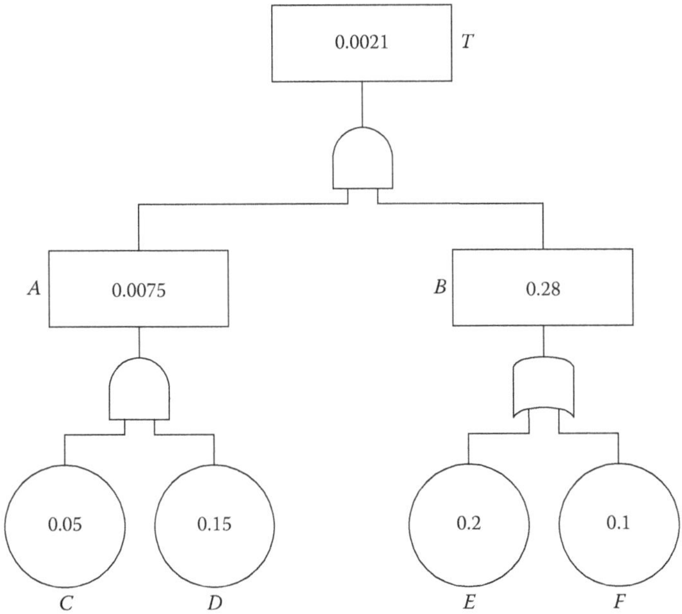


**FIGURE 9.4**
A fault tree with probability values for the given and calculated fault event occurrence.

Using the above-calculated values and Chapter 4 material concerning fault tree analysis, we get

$$\begin{aligned} P(T) &= P(A)\,P(B) \\ &= (0.0075)\,(0.28) \\ &= 0.0021 \end{aligned} \tag{9.3}$$

where
$P(T)$ is the probability of occurrence of event $T$.

Thus, the probability of occurrence of the top event $T$ (i.e., a robot accident involving a human) is 0.0021. The fault tree in Figure 9.3 with given and calculated probability values for fault event occurrence is shown in Figure 9.4.

### 9.8.5 Markov Method

This is a quite useful method and, over the years, has been used to perform analyses of safety and reliability of human–machine systems [17]. The method is described in Chapter 4 and its application to various types of human–machine systems are demonstrated in [17].

The application of the Markov method to robot-related accidents in workplaces is demonstrated through the example below.

**EXAMPLE 9.2**

Assume that a robot used in a workplace can either fail due to an accident involving a human or fail safely. The robot failure rate due to an accident involving a human is $\lambda_a$ and its safe failure rate is $\lambda_s$. The state-space diagram describing this scenario is shown in Figure 9.5. The numerals in the circles and box denote robot states.

Develop expressions for the robot state probabilities by using the Markov method and assuming that the robot failures occur independently and its failure rates are constant.

The following symbols are associated with the Figure 9.5 state-space diagram:

$i$ is the $i$th state of the robot, where $i = 0$ (robot operating normally), $i = 1$ (robot failed due to an accident involving a human), $i = 2$ (robot failed safely).
$P_i(t)$ is the probability that the robot is in state $i$ at time $t$, for $i = 0,1,2$.
$\lambda_s$ is the constant failure rate of the robot failing safely.
$\lambda_a$ is the constant failure rate of the robot failing due to an accident involving a human.

Using the Markov method described in Chapter 4 and Figure 9.5, we write down the following equations:

$$\frac{dP_0(t)}{dt} + (\lambda_a + \lambda_s)P_0(t) = 0 \tag{9.4}$$

$$\frac{dP_1(t)}{dt} - \lambda_a P_1(t) = 0 \tag{9.5}$$

$$\frac{dP_2(t)}{dt} - \lambda_s P_2(t) = 0 \tag{9.6}$$

At time $t = 0$, $P_0(0) = 1$, $P_1(0) = 0$, and $P_2(0) = 0$.

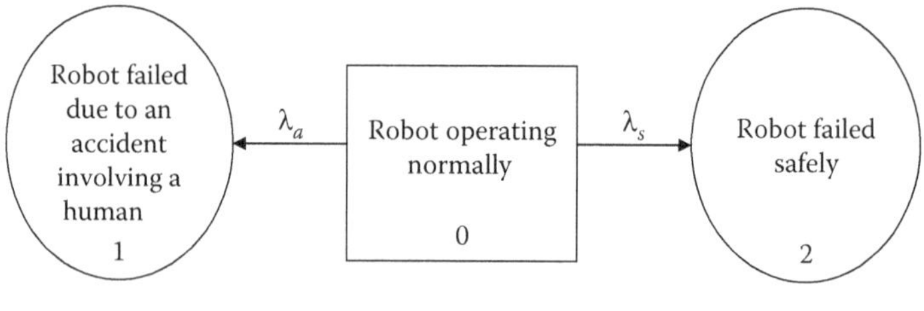


**FIGURE 9.5**
State-space diagram of a robot used in a workplace.

By solving Equations 9.4 through 9.6, we get

$$P_0(t) = e^{-(\lambda_a+\lambda_s)t} \tag{9.7}$$

$$P_1(t) = \frac{\lambda_a}{\lambda_a + \lambda_s}[1 - e^{-(\lambda_a+\lambda_s)t}] \tag{9.8}$$

$$P_2(t) = \frac{\lambda_s}{\lambda_a + \lambda_s}[1 - e^{-(\lambda_a+\lambda_s)t}] \tag{9.9}$$

**EXAMPLE 9.3**

Assume that the constant failure rate of a robot used in a workplace and failing safely is 0.05 failures per hour and its constant failure rate failing due to an accident involving a human is 0.01 failures per hour. Calculate with the aid of Equations 9.7 through 9.9, the probabilities of the robot being in states 0, 1, and 2 during a 100 h mission.

By substituting the given data values into Equations 9.7 through 9.9, we obtain

$$\begin{aligned} P_0(100) &= e^{-(0.01+0.05)(100)} \\ &= 0.0024 \end{aligned}$$

$$\begin{aligned} P_1(100) &= \frac{0.01}{(0.01+0.05)}[1 - e^{(0.01+0.05)(100)}] \\ &= 0.1662 \end{aligned}$$

and

$$\begin{aligned} P_2(100) &= \frac{0.05}{(0.01+0.05)}[1 - e^{-(0.01+0.05)(100)}] \\ &= 0.8313 \end{aligned}$$

Thus, the probabilities of the robot being in states 0, 1, and 2 during a 100 h mission are 0.0024, 0.1662, and 0.8313, respectively.

## 9.9 Problems

1. What are the primary human factors-related issues during the robotic systems' factory integration process?
2. Discuss in detail four issues of the above question.

3. What are the common robot tasks?
4. List and discuss common robot-related human tasks.
5. What are the rules of robotics with regard to humans?
6. What are the advantages and disadvantages of robotization with respect to humans?
7. What are the risk-reducing measures to prevent robot-related human accidents?
8. Discuss at least six useful guidelines to safeguard robot teachers.
9. What are the robot movement limitation approaches? Discuss each of these approaches in detail.
10. Describe the following methods with regard to robotics workplaces:
    a. Energy barrier analysis
    b. Deviation analysis
    c. Fault tree analysis

## References

1. Parsons, H.M., Human factors loom as vital issues in robotics, *Robotics Today*, 1987, pp. 27–31.
2. Howard, J.M., Focus on the human factors in applying robotic systems, *Robotics Today*, 1982, pp. 32–34.
3. Dhillon, B.S., *Robot Reliability and Safety*, Springer-Verlag, New York, 1991.
4. Helander, M.G., Ergonomics and safety considerations in the design of robotics workplaces: A review and some priorities for research, *International Journal of Industrial Ergonomics*, Vol. 6, 1990, pp. 127–149.
5. American National Standard for Industrial Robots and Robot Systems—Safety Requirements, *Document No. ANSI/RIA R15.06*, 1986; available from the Robotic Industries Association (RIA), 900 Victors Way, P.O. Box 3724, Ann Arbor, Michigan.
6. Yokomizo, Y., Hasegawa, Y., Komatsubara, A., Problems of, and industrial medicine measures for, the introduction of robots, *Occupational Health and Safety in Automation and Robotics* ed. K. Noro, Taylor & Francis, London, 1987, pp. 309–325.
7. Owen, T., *Assembly with Robots*, Kogan Page, London, 1985.
8. Carlsson, J., *Robot Accidents in Sweden, Report No. 1984.2*, Arbetarskyddsstgrelsen, Stockholm, Sweden, 1984.
9. Dhillon, B.S., *Engineering Safety: Fundamentals, Techniques, and Applications*, World Scientific Publishing, River Edge, New Jersey, 2003.
10. Harms-Ringdahl, L., Experiences from safety analysis of automatic equipment, *Journal of Occupational Accidents*, Vol. 8, 1986, pp. 139–148.

11. Rahimi, M., Systems safety for robots: An energy barrier analysis, *Journal of Occupational Accidents*, Vol. 8, 1986, pp. 127–138.
12. Umezaki, S., Sugimoto, N., A study on safety evaluation index and industrial accident analysis from the viewpoint of the safety confirmation type, in *Ergonomics of Hybrid Automated Systems I*, ed. H.R. Parsaei and M.R. Wilhelm, Elsevier, Amsterdam, The Netherland, 1988, pp. 491–498.
13. Kirwan, B., Ainsworth, L.K., *A Guide to Task Analysis*, Taylor & Francis, London, 1992.
14. *Guidelines for Controlling Hazardous Energy*, National Institute for Occupational Safety and Health (NIOSH), Morgan Town, West Virginia, 1983.
15. Dhillon, B.S., *Design Reliability: Fundamentals and Applications*, CRC Press, Boca Raton, Florida, 1999.
16. Dhillon, B.S., Singh, C., *Engineering Reliability: New Techniques and Applications*, John Wiley and Sons, New York, 1981.
17. Dhillon, B.S., *Human Reliability: With Human Factors*, Pergamon Press, New York, 1986.

# 10

# *Robot Testing, Costing, and Failure Data*

## 10.1 Introduction

Robot testing is very important because it plays an important role in the success or failure of robotization. Two common types of robot testing are known as robot performance testing and robot reliability testing [1]. Over the years, a number of methods have been developed to conduct performance testing; the forms of performance testing vary from the verification of design goals by robot manufacturers to the determination of the most suitable robot for a specific application. There are various types of robot reliability testing and their purpose is to obtain knowledge of robot failures.

Robot costing is very important as cost plays a crucial role, directly or indirectly, in robotization in the industrial sector. For example, the economic viability of robot use in the factory depends on factors such as investment cost (i.e., acquisition cost, installation cost, and cost of adaptation to the workshop), ownership cost (e.g., maintenance cost and fuel cost), and return on investment [2].

Robot failure data provides invaluable information to concerned professionals and management alike as it indicates the results of the reliability-related effort put in during the design and manufacture phases of robots. Today, there are various sources for obtaining data related to robot failures.

This chapter presents various important aspects of robot testing, costing, and failure data.

## 10.2 Robot Performance Testing

There are several forms of robot performance testing. Two examples of such forms are the verification of design goals by the robot manufacturer and the determination of the most suitable robot for a specific application. Three performance parameters that generally play a key role in robot performance testing are shown in Figure 10.1 [1,3].

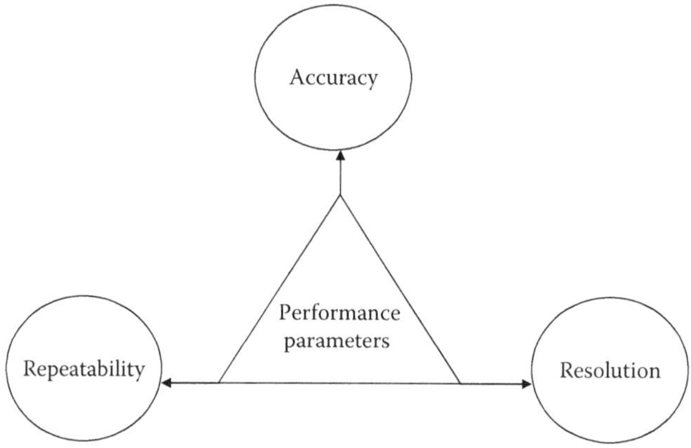


**FIGURE 10.1**
Performance parameters that play a key role in robot performance testing.

Accuracy is concerned with measuring the differential between the directions given to a robot regarding movement(s) and its actual movements. Repeatability may simply be described as the closeness of concurrence of a robot's repeated movements in the same location when subject to the same conditions [1,4]. It is to be noted that repeatability is one of the greatest advantages of a robot because of its ability to repeat programmed motions consistently.

Finally, resolution, in regard to robots, is determined by the capability of feedback devices such as encoders and resolvers for determining the locality of a specific point and, ultimately, the end point position of the robot arm. Nonetheless, in broader perspectives, resolution may simply be stated as a measure of the change in variable output of a device.

It is to be noted that most of the industrial robot performance tests may be tailored to measure either of two types of behavior: path and point-to-point. Parameters such as kinematic repeatability or kinematic accuracy are measured by path tests as opposed to dynamic stability, quasi-static repeatability, or quasi-static accuracy by point-to-point tests.

Some of the general guidelines concerned with industrial robot performance testing are presented below [1,5]:

- Ensure that the appropriate safety rules are properly followed during the robot testing and programming process.
- Ensure that the involved test equipment is temperature- and time-stable.
- Ensure that for tests with multiple permutations, only one control parameter at a time is changed, because parameters such as repeatability depend upon the robot's speed, payload, temperature, and so on.

- Ensure the proper rigidity of the mounting platform of the robot.
- Ensure that noncontact gauges such as laser interferometers, analog distance sensors, and vision systems are employed instead of contact gauges (e.g., dial indicators).
- Ensure that the measuring equipment resolution is at least one order of magnitude greater than the actual value to be measured.

## 10.3 Robot Performance Testing Methods

Over the years, many test methods have been developed to evaluate the performance of robots [5]. Four of these test methods are described below, separately.

### 10.3.1 Test-Robot Program Method

The test-robot program method is concerned with measuring the position and orientation errors acquired during static positioning, and are computed statistically as three-dimensional error vectors. The method has two versions (i.e., Mark I and Mark II) developed by the same individual [5,6]. Some of the important features of this method are as follows [1,5,6]:

- It possesses the capability to be employed as a simulation tool to evaluate alternative solutions.
- It possesses the user-friendly and run-time-checked screen management system.
- It possesses the capability to undertake analysis for hand position and orientation error for industrial robots.
- It possesses the capability to save end results in a user-specified text file on disk.
- It possesses the capability for evaluating perpendicularity errors of the used test rig for robots.
- It possesses the capability for disk output/input error handling and run-time input line editing.

### 10.3.2 Ford Method

The Ford method was developed by the Ford Motor Company. It describes industry-oriented approaches to certify industrial robots using the Selspot, Inc., vision system [1]. The method calls for putting the robot in question under a cycling test consisting of the movement of all axes from one end of motion to the other at full-rated payload and at speed for about 40 h continuously.

This is basically a noninstrumented test and any duty cycle-related limitation is followed closely. The two objectives of the test are as follows [1]:

- Obtain information concerning the integrity and reliability of the robot system
- Find infant mortality faults such as usage of undersized fuses and loose bolts

Additional information on this method is available in [1].

### 10.3.3 IPA–Stuttgard Method

The IPA–Stuttgard method illustrates a general procedure to measure static positioning errors and dynamic path parameters. It was developed at the Institute for Production and Automation (IPA), Stuttgart, Germany [1,5,7]. Two main requirements considered during the development of this robot performance testing method are as follows [1,7]:

- *Requirement I: Highly automated test bed.* The main idea behind this requirement is to lower the time required to evaluate the performance of a robot.
- *Requirement II: Exact coordination between the measuring equipment used and the robot.* The main reason behind this requirement is to permit the measurement of characteristics within the work zone of a robot with an acceptable level of accuracy.

Additional information on this method is available in [7].

### 10.3.4 National Bureau of Standards Method

The National Bureau of Standards (NBS) of the United States developed this robot performance test method to measure the performance of industrial robots. The NBS method uses laser-based instrumentation [1,8]. The basis for the method is the idea of continuously controlling the angles of projection of a laser beam, so that the beam will impinge on the target mirror installed on the robot's wrist.

Additional information on the method is available in [5].

## 10.4 Robot Reliability Testing

The basic objective of robot reliability testing is to gain knowledge about robot failures, as this knowledge plays an important role in controlling

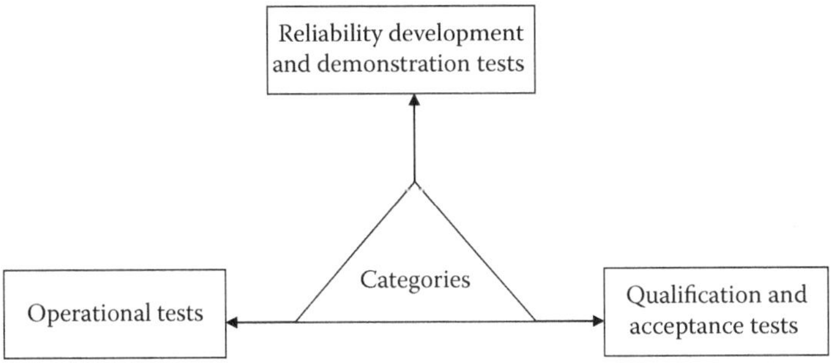


**FIGURE 10.2**
Categories of robot reliability tests.

failure tendencies and their effects. Robot reliability tests may be grouped under three broad categories as shown in Figure 10.2 [1,9].

The main objectives of the reliability development tests and their demonstration are to verify improvements in the reliability of the design, to determine if the robot design has to be improved to meet the specified reliability requirements, and to point out any required design changes. Two objectives of the qualification and acceptance tests are to determine if a specific design is qualified for its proposed use and to determine if an item is acceptable or not acceptable.

Finally, the objectives of operational tests are to verify the reliability analyses performed during the design stage, to provide the necessary data to justify required operational procedure and operational changes with regard to robot reliability and maintainability, and to provide information that would be useful in the activities to be followed.

### 10.4.1 Management-Related Tasks in a Robot Reliability Test Plan

A typical robot test plan is composed of many tasks that generally require frequent updating. The plan includes tasks such as the following [1,9]:

- Allocate time and money.
- Determine test needs and objectives.
- Review current or planned functional tests and operations from the standpoint of determining if certain required information could be extracted from such tests and operations.
- Formulate individual test needs.
- Evaluate the already available data for meeting any existing needs, thus eliminating the requirements for specific tests.
- Determine the types of tests to be performed.
- Review and approve test reporting methods and documents.
- Keep information on status of the test.

### 10.4.2 Useful Guidelines for Robot Reliability Demonstration Tests

Some of the useful guidelines to perform robot reliability demonstration tests are as follows [1,9]:

- Determine the test duration by considering factors such as the availability of the robot equipment to be tested, test facilities, and testing manpower.
- Develop a comprehensive description of the robot equipment with respect to factors such as operational environment, mission time, service life, and operational mission.
- Prepare a list of the robot system items that are to be excluded from the demonstration test under consideration.
- Pay attention to developing appropriate test monitoring approaches whether manual or automatic.
- Put in an appropriate amount of effort to relate test environments to the expected robot use environments.
- Pay attention to choosing the methods to conduct data analysis.
- Pay attention to choosing the test plan.

## 10.5 Robot Testing and Start-Up Safety-Related Factors

During robot testing and start-up, there are many safety-related factors that have to be considered to improve robot safety. Some of these factors are as follows [1,10]:

- Following the instructions of robot manufacturers for testing and repair.
- After each robot hardware or software modification, maintenance, and repair, a procedure for the restart of the robot should incorporate steps such as function testing the robot for the correct system operation and checking, prior to turning on the power, for any changes or additions to the hardware of the robot.
- Ensuring that no human is present in the work zone of the robot until the safeguarding and the appropriate operation verification is completed.
- The proper incorporation of two verification procedures—that is, I and II. Verification procedure I is concerned with the verification of the installation of various items as required, before turning on the power. Some examples of these items are electrical

connections, limiting devices to restrict the work zone, peripheral equipment and systems, utility connections, mechanical mounting and stability, and communication connections. Similarly, verification procedure II, after turning on the power, is concerned with the verification of items such as the proper functioning of safeguards, interlocks, emergency stop devices, automatic operations as specified, movement of each axis and its restriction as intended, slow speed as specified in [10], and program execution as specified.

## 10.6 Robot Project Cost

The project cost of robotization may be divided into the following eight categories [1,11]:

- *Category I: Computer cost.* This comprises the hardware and software costs. The hardware cost is associated with items such as the central processor unit, interface modules, and core store. The software cost is associated with items such as customized software and standard software packages.
- *Category II: Safety cost.* This cost accounts for around 10% of the basic system cost and is composed of elements such as the cost of protection devices for the operator(s), of alarm devices, safety enclosure, and safety training.
- *Category III: Manufacturing equipment cost.* This cost is associated with items such as the robot system (this includes the manipulator and associated control system and motive source), grippers or workhandlers, system interfaces, and package enhancements.
- *Category IV: Project management cost.* This cost is associated with managing the project and includes items such as the salary of the project manager and cost of secretarial help.
- *Category V: System engineering cost.* In this cost, labor is the dominant factor because the time of technicians and engineers is quite costly. Labor time is associated with tasks such as purchasing equipment for the system, testing equipment against specifications of manufacturers, testing the system against system criteria, reassembly, and retesting.
- *Category VI: Ancillary equipment and structural modification cost.* This cost is concerned with the work to be carried out prior to the robot installation.

- *Category VII: Cost associated with personnel.* This cost is associated with items such as skills training, redundancies, redeployment, recruitment, and wage settlements.
- *Category VIII: Miscellaneous cost.* This cost includes the costs of items such as feasibility studies, purchase of spares, and maintenance costs.

## 10.7 Models for Estimating Robot-Related Costs

Over the years, many mathematical models have been developed to estimate various types of robot-related costs. Some of these models are presented below [1].

### 10.7.1 Model I

This model is concerned with estimating the robot operating cost. The cost is defined by

$$ROC = DLC + RMC + SPFC + MC + ATCC \tag{10.1}$$

where
  $ROC$ is the robot operating cost.
  $DLC$ is the direct labor cost to tend the robot.
  $RMC$ is the cost associated with the robot related maintenance activities.
  $SPFC$ is the cost associated with supplying parts to feeders.
  $MC$ is the miscellaneous cost.
  $ATCC$ is the cost of adjusting tools for changeovers.

### 10.7.2 Model II

This model is concerned with estimating the investment cost associated with a robotization project. The cost is expressed by

$$RPIC = RC_p + RC_e + RC_{st} + RC_i + RC_m \tag{10.2}$$

where
  $RPIC$ is the investment cost of the robotization project.
  $RC_p$ is the robot procurement cost.
  $RC_e$ is the robot engineering cost.
  $RC_{st}$ is the special tooling cost of the robot.

$RC_i$ is the robot installation cost.
$RC_m$ is the miscellaneous cost of the robotization project.

### 10.7.3 Model III

This model is concerned with estimating the annual corrective maintenance cost of a robot. The cost is defined by

$$RCMC_a = (LC_m)(OH_{as})\left[\frac{MTTR_r}{MTBF_r}\right] \tag{10.3}$$

where
$RCMC_a$ is the annual corrective maintenance cost of the robot.
$OH_{as}$ is the annual scheduled operating hours of the robot.
$LC_m$ is the robot maintenance labor cost per hour.
$MTTR_r$ is the mean time to repair of the robot.
$MTBF_r$ is the mean time between robot failures.

**EXAMPLE 10.1**

Assume that the annual scheduled operating hours of a robot are 4000 h and its mean time between failures is 1200 h. The mean time to repair of the robot is 8 h and maintenance labor cost is US$30 per hour.

Calculate the annual corrective maintenance cost of the robot.

By substituting the specified data values into Equation 10.3, we get

$$RCMC_a = (30)(4000)\left[\frac{8}{1200}\right]$$
$$= \$800$$

Thus, the annual corrective maintenance cost of the robot is US$800.

### 10.7.4 Model IV

This model is concerned with estimating the unit cost of assembly by the robot. The cost is expressed by [12]

$$C_u = T(C_1 + C_2) \tag{10.4}$$

where
$C_u$ is the unit cost of assembly by the robot.
$T$ is the time taken to assemble the product expressed in seconds.
$C_1$ is the per-second cost of the flexible portions of the flexible assembly system.
$C_2$ is the per-second cost of the dedicated portions of the flexible assembly system.

### 10.7.5 Model V

This model is concerned with estimating the possible cost of losses to a robot manufacturer from a robot accident. The cost is expressed by

$$PC_{rf} = C_{ai} + C_{pc} + C_{pm} + C_s + C_{pd} + C_{ii} + C_{lf} + C_m \tag{10.5}$$

where

$PC_{rf}$ is the possible cost of losses to a robot manufacturer from a robot-related accident.
$C_{ai}$ is the cost of accident investigation.
$C_{pc}$ is the cost of loss in public confidence.
$C_{pm}$ is the cost associated with preventive measures to stop recurrences.
$C_s$ is the settlement cost of death or injury claims.
$C_{pd}$ is the cost associated with property damage claims not covered by insurance.
$C_{ii}$ is the cost of increment in insurance.
$C_{lf}$ is the cost of legal fees for defense against claims.
$C_m$ is the miscellaneous cost.

### 10.7.6 Model VI

This model is concerned with estimating the robotic installation and start-up cost. The cost is expressed by [13]

$$RISC = AC + PC + TC + EC + IC + ACCS + REC + TRC \tag{10.6}$$

where

$RISC$ is the robotic installation and start-up cost.
$AC$ is the acquisition cost, which includes costs of the robot and controller.
$PC$ is the programming cost.
$TC$ is the tooling cost, which includes costs of grippers, fixtures, and end-effectors.
$EC$ is the engineering cost.
$IC$ is the installation cost, which includes costs of set-up labor, utility hook-up and foundations.
$ACCS$ is the cost of accessories; these accessories include items such testers, part feeders, bins, and conveyors.
$REC$ is the cost of related expenses (e.g., insurance).
$TRC$ is the training cost.

## 10.8 Robot Life Cycle Cost Estimation Models

There are a number of life cycle cost models that can be used to estimate the life cycle cost of robots [14,15]. Two of such models are presented below [14,15].

### 10.8.1 Life Cycle Cost Model I

This model assumes that the cost of the robotic life cycle is made up of three major components: failure cost, operating cost, and initial cost. Thus, the robot life cycle cost is expressed by [1]

$$LCC_r = C_{ri} + C_{rf} + C_{ro} \tag{10.7}$$

where

- $LCC_r$ is the cost of the robot's life cycle.
- $C_{ri}$ is the robot's initial cost, which is made up of the cost of components such as initial spares, of training, equipment, and software.
- $C_{rf}$ is the robot failure cost, which is made up of components such as repair cost and the cost of operational losses attributable to robot-related failures.
- $C_{ro}$ is the robot operating cost, which is made up of elements such as salaries, cost of expendables, and preventive maintenance cost.

### 10.8.2 Life Cycle Cost Model II

This model assumes that the robot's life cycle cost is made up of two major components: nonrecurring cost and recurring cost. Thus, the robot's life cycle cost is defined by

$$LCC_r = C_{nr} + C_r \tag{10.8}$$

where

- $LCC_r$ is the robot's life cycle cost.
- $C_{nr}$ is the robot nonrecurring cost.
- $C_r$ is the robot recurring cost.

The robot nonrecurring cost is made of elements such as listed below:

- Robot cost
- Engineering cost
- Installation cost
- Training cost
- Reliability/maintainability improvement cost

The robot's cost is made up of the cost of items such as test equipment, material handling equipment, tooling for the robot hand, and cost of robot modifications. The engineering cost is concerned with items such as system planning, layout design, programming the robot, and product redesign. The installation cost includes the cost of items such as safety equipment, vendor installation assistance, building modifications, and relocating other equipment.

The training cost is associated with items such as training of operators, programmers, and maintenance-related personnel. Finally, reliability/maintainability improvement cost is associated with the actions concerning robot-related reliability and maintainability improvement.

Some of the main elements of the recurring costs of robots are as follows:

- Operating cost
- Manpower cost
- Maintenance cost
- Inventory cost

## 10.9 Useful Methods for Making Financial Decisions about Robotization

There are numerous methods that can be used to make investment decisions regarding robotization [1,14]. Some of these methods are presented below [1].

### 10.9.1 Method I: Life Cycle Costing

This method calls for estimating the life cycle cost of the alternative projects and then comparing the end results and selecting the project with the lowest life cycle cost. The life cycle cost of a project is expressed by

$$PLC = \sum_{j=0}^{n}\left[\frac{AC_j + RC_j}{(1+i)^j}\right] - \frac{SV}{(1+i)^n} \tag{10.9}$$

where

$PLC$ is the project life cycle cost.
$n$ is the project life in years.
$i$ is the annual interest rate.
$RC_j$ is the estimated running cost in year $j$.
$AC_j$ is the acquisition cost in year $j$.
$SV$ is the estimated salvage value of the project under consideration.

### 10.9.2 Method II: Payback Period

This method is concerned with determining the length of time needed for recovering the cost of an investment. After the payback period, the robot will begin making positive returns for its entire remaining life. Thus, the robot installation payback method is expressed by [1,16]

$$N_p = \frac{I_t}{LS_t - E_{rt}} \tag{10.10}$$

where

$N_p$ is the total number of years for payback of the investment cost.
$I_t$ is the total investment for the robot and its associated accessories.
$LS_t$ is the total annual labor savings.
$E_{rt}$ is the total annual expense of the robot.

It is to be noted that in the payback analysis method, the robot production rate's effects can also be considered. Furthermore, in situations where costly production equipment is used, the payback is more sensitive to the production rate.

Thus, the robot payback period is defined by

$$N_{rp} = \frac{I_t}{LS_t - E_{rt} \pm \theta(LS_t + DC_{ae})} \tag{10.11}$$

where

$N_{rp}$ is the total number of years for the payback on the investment in the robot.
$\theta$ is the production rate coefficient.
$DC_{ae}$ is the annual depreciation cost of associated equipment.

**EXAMPLE 10.2**

Assume that the total amount of money invested for a robot and its accessories is US$100,000. The annual cost of the robot's upkeep and the expected annual labor savings are US$20,000 and US$30,000, respectively.

Calculate the payback period for the investment in the robot.

By substituting the specified data values into Equation 10.10, we get

$$N_p = \frac{100{,}000}{30{,}000 - 20{,}000} = 10 \text{ years}$$

Thus, the payback period for the investment in the robot is 10 years.

### 10.9.3 Method III: Minimum Cost Rule

In this method, costs such as capital cost and operating cost are calculated for each and every alternative, and a decision on the desirability of the investment is taken by comparing the cost totals of the alternatives. An example of the minimum cost rule method is the present value approach.

The net present value of a robot project under consideration can be estimated with the aid of the equation presented below [1,13]:

$$PV_n = \sum_{j=0}^{n} \frac{(PS_j - AC_j - RC_j)}{(1+i)^j} + \frac{V_s}{(1+i)^n} \tag{10.12}$$

where

$PV_n$ is the net present value of the robotization project under consideration.
$V_s$ is the expected salvage value of the robotization project under consideration.
$i$ is the annual interest rate.
$n$ is the robot's project life in years.
$AC_j$ is the acquisition cost in year $j$.
$RC_j$ is the expected running cost in year $j$.
$PS_j$ is the expected potential savings in year $j$.

It is to be noted that during the alternative projects' comparing process, the maximum present value is aimed for.

### 10.9.4 Method IV: Return on Investment

This is probably the most celebrated method to compare alternative investments under consideration. In the case of a robotization project, once the cost of capital for investment is established, then it should be compared with the return on investment.

The return on investment is defined by [1]

$$I_r = \left(\frac{S_t}{I_{ra}}\right)(100\%) \tag{10.13}$$

where

$I_r$ is the return on investment.
$S_t$ is the total annual savings resulting from the robot.
$I_{ra}$ is the total investment in the robot and its associated accessories.

## 10.10 Failure Data Uses with Regard to Robots and Failure Reporting and Documentation System for Robots

In various robot-related studies, failure-related data plays an important role. More specifically, some of the areas in which data become very useful are as follows [11]:

- Gaining knowledge concerning the robot's design and manufacturing-related shortcomings
- Estimating availability, reliability, and hazard rate (failure rate) of the robot
- Recommending design changes to improve the reliability of the robot or of its parts
- Supporting reliability growth-related programs for the robot
- Performing reliability cost trade-off studies for the robot
- Determining the maintenance needs of a robot and its parts
- Conducting robot design reviews
- Conducting replacement-related studies for the robot
- Performing life cycle cost-related studies for the robot

The establishment of proper failure reporting and documentation systems is essential to produce reliable and safe robots. During the designing process of such systems, it is important to pay proper attention to factors such as shown in Figure 10.3 [1].

## 10.11 Main Data Sources for Reliability

There are many sources for obtaining data related to product reliability. These include previous experience with similar products, tests (environmental qualification, field installation, and field demonstration), repair records, warranty claims, inspection records, and so on. In the case of robots, as they may contain electrical, mechanical, and electronic parts; the failure data for such parts may be obtained from the existing failure data banks in various industrial sectors.

An excellent document to estimate the failure rates of electronic parts is *MIL-HDBK-217E* [17]. Nonetheless, the main data sources for reliability are as follows [1,18]:

- *Laboratory tests.* In this case, under laboratory conditions, a group of similar items is tested and the failure events are recorded along with the time intervals.
- *Published literature (information).* This involves obtaining generic data for failure from open published literature (e.g., government publications, industrial reports, journals, books, and conference proceedings). It is to be noted that the data published in such documents is

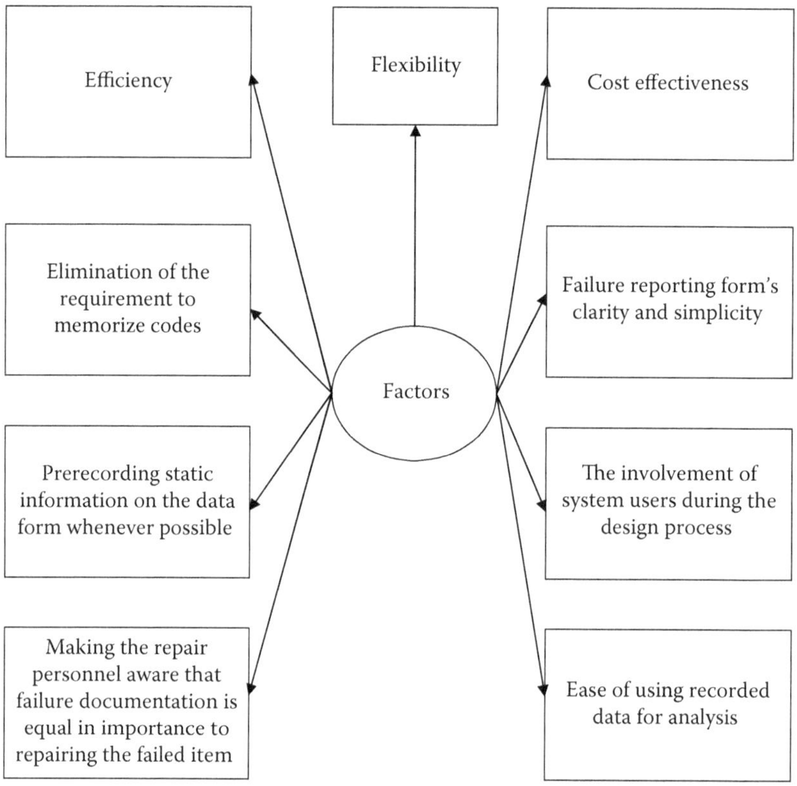


**FIGURE 10.3**
Factors requiring attention in designing failure reporting and documentation systems for robots.

generally in the form of failure rates, and the information on population and experience is missing; thus, it lowers the usefulness of such data.

- *Expert opinions.* In this case, data on failure is generated from the opinions expressed by experts such as equipment maintenance personnel and operators. Subsequently, their judgments are converted into quantified data with the aid of methods such as Delphi and paired comparisons.
- *Event data banks.* Specifically designed systems that are capable of accepting the details of all types of events and their occurrence times, are normally referred to as "event data banks." Data on failure and other plant management-related information is extracted by processing the available event records.

- *Field failure data collection.* This is the approach of making use of the information available in operating and maintenance records for establishing an as-complete-as-possible historical record of events. From this history, data on failure is derived.

## 10.12 Repair and Inspection Records-Related Requirements for Robots

Each time a robot is repaired or inspected, various types of information can be recorded. This information becomes very useful to make robot-related decisions. Robot repair records should include information on the following items at least [1,19]:

- Serial number
- Robot designation
- Name(s) of the repair person(s)
- Repair starting time and date
- Repair completion time and date
- Operating time since the previous repair
- Model number
- Repair description and the place of repair
- Type of inspection that resulted in the repair action

Similarly, the periodic inspection records should include information on the following items at least [1,19]:

- Serial number and model
- Date of inspection
- Inspector identification
- Robot designation
- Inspection location
- Inspection results
- Inspection method used
- Action to be performed for the faulty portion
- Action implemented and implementation date

Information on some possible documents, banks, and organizations to obtain data on failure for robot-related reliability studies is available in [1].

## 10.13 Problems

1. Describe performance parameters that play a key role in robot performance testing.
2. Describe the following robot performance testing methods:
   a. The test-robot program method
   b. The Ford method
3. What are the three broad categories of robot reliability tests? Discuss each of these categories.
4. What are the management-related tasks in a robot reliability test plan?
5. What are the robot testing and start-up safety-related factors?
6. What are the main categories of the robot project cost?
7. Assume that a robot's mean time to failure is 1600 h and its annual scheduled operating hours are 4800 h. The robot's mean time to repair is 10 h and its maintenance labor cost is US$40 per hour. Calculate the robot's annual corrective maintenance cost.
8. What are the factors that require proper attention when designing failure reporting and documentation systems for robots?
9. What are the main data sources for reliability? Discuss each of these sources.
10. List at least seven useful guidelines for robot reliability demonstration tests.

## References

1. Dhillon, B.S., *Robot Reliability and Safety*, Springer-Verlag, New York, 1991.
2. Nof, S.Y., Decisions aids for planning industrial robot operations, *Proceedings of the Annual Industrial Engineering Conference*, 1982, pp. 180–185.
3. Wodzinski, M., Putting robots to the test, *Robotics Today*, 1987, pp. 17–20.
4. Robotics Industries Association, *Robotics Glossary*, Ann Arbor, Michigan, 1984.

5. Ranky, P.G., Wodzinski, M., *A Survey of Robot Test Methods with Examples, Report No. RSD-TR-1-87*, Center for Research on Integrated Manufacturing, Robot Systems Division, College of Engineering, University of Michigan, Ann Arbor, Michigan, February 1987.
6. Ranky, P.G., Ho, C.Y., *Robot Modeling: Control and Applications with Software*, Springer-Verlag, Berlin, 1985.
7. Warnecke, H.J., Test stand for industrial robots, *Proceedings of the 7th International Symposium on Industrial Robotics*, October 1977, pp. 10–15.
8. Lau, K., Hocken, R., Haight, W., *An Automatic Laser-Tracking Interferometer System for Robot Metrology, Report*, National Bureau of Standards, Gaithersburg, Maryland, March 1985.
9. Von Alven, W.H., Ed., *Reliability Engineering*, Prentice-Hall, Englewood Cliffs, New Jersey, 1964.
10. American National Standards Institute, *American National Standard for Industrial Robots and Robot Systems: Safety Requirements, ANSI/RIA R15.06, 1986*; available from the American National Standards Institute (ANSI), 1430 Broadway, New York, New York 10018, 1986.
11. Morgan, C., *Robots Planning and Implementation*, Springer-Verlag, Berlin, 1984.
12. Scott, P.B., Husband, T.M., Robotic assembly: Design, analysis and economic evaluation, *Proceedings of the 13th International Symposium on Industrial Robots*, 1983, pp. 5.12–5.29.
13. Hasegawa, Y., Evaluation and economic justification, in *Handbook of Industrial Robotics*, Ed. S.Y. Nof, John Wiley and Sons, New York, 1985, pp. 665–687.
14. Dhillon, B.S., *Life Cycle Costing for Engineers*, CRC Press, Boca Raton, Florida, 2010.
15. Dhillon, B.S., *Life Cycle Costing: Techniques, Models, and Applications*, Gordon and Breach Science Publishers, New York, 1989.
16. Miller, R.K., *Industrial Robot Handbook*, SEAI Institute, P.O. Box 590, Madison, Georgia, 1983.
17. United States Department of Defense, *MIL-HDBK-217E, Reliability Prediction of Electronic Equipment*, Washington, D.C., October 1986; available from the Naval Publications and Forms Center, 5801 Tabor Avenue, Philadelphia, Pennsylvania 19120, USA.
18. Bello, G.C., Data validation procedures, in *Reliability Data Bases*, eds. A. Amendola and A.Z. Keller, Reidel, Boston, 1987, pp. 125–132.
19. Japanese Industrial Safety and Health Association, *An Interpretation of the Technical Guidance on Safety Standards in the Use, Etc., of Industrial Robots*, 3-35-1, Shiba, Minato-Ku, Tokyo 108, Japan, 1985.

# 11

# *Mathematical Models for Analysis of Robot-Related Reliability and Safety*

## 11.1 Introduction

Mathematical modeling is a commonly used approach to conduct various types of analysis in the area of engineering. In this case, the components of a system are represented by idealized elements assumed to have all the representative characteristics of real-life components, whose behavior is possible to be described by equations. However, it is to be noted that the degree of realism of a mathematical model depends on the type of assumptions imposed upon it.

Over the years, many mathematical models have been developed to analyze robot-related reliability and safety. Many of these models were developed using the Markov method [1,2]. Although the effectiveness of such models can vary quite considerably from one application area to another, some of them are being used quite successfully to study various types of real-life problems in industries using robot systems.

This chapter presents a number of mathematical models considered useful to perform various types of analysis of the reliability and safety of robotic systems. All of these models are based upon the Markov method.

## 11.2 Model I

This mathematical model represents a robot having three distinct states: working normally, failed due to a hardware or software failure, and failed due to a human error. The failed robot is repaired to revert to its working state from both the failed states. The robot state-space diagram is shown in Figure 11.1. The numerals in rectangles denote system states.

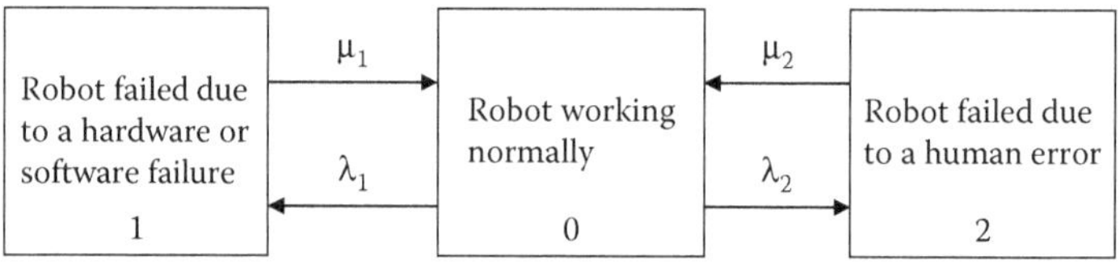


**FIGURE 11.1**
Robot state-space diagram.

The following four assumptions are associated with this mathematical model [1,3]:

- Robot hardware or software failure and human error rates are constant.
- The failed robot repair rates are constant.
- The repaired robot is as good as new.
- Human error and hardware or software failures are statistically independent.

The following symbols are associated with the diagram shown in Figure 11.1 and its associated equations:

$P_j(t)$ is the probability that the robot is in state $j$ at time $t$; for $j = 0$ (working normally), $j = 1$ (failed due to a hardware or software failure), $j = 2$ (failed due to a human error).
$\lambda_1$ is the robot hardware or software failure rate.
$\lambda_2$ is the robot human error rate.
$\mu_1$ is the robot repair rate from failed state 1.
$\mu_2$ is the robot repair rate from failed state 2.

Using the Markov method, we have the following equations for the state-space diagram shown in Figure 11.1 [1,3]:

$$\frac{dP_0(t)}{dt} + (\lambda_1 + \lambda_2)P_0(t) = \mu_2 P_2(t) + \mu_1 P_1(t) \tag{11.1}$$

$$\frac{dP_1(t)}{dt} + \mu_1 P_1(t) = \lambda_1 P_0(t) \tag{11.2}$$

$$\frac{dP_2(t)}{dt} + \mu_2 P_2(t) = \lambda_2 P_0(t) \tag{11.3}$$

At time $t = 0$, $P_0(0) = 1$, $P_1(0) = 0$, and $P_2(0) = 0$.

Solving Equations 11.1 through 11.3 using Laplace transforms, we obtain

$$P_0(t) = \frac{\mu_1\mu_2}{n_1 n_2} + \left[\frac{(n_1+\mu_1)(n_1+\mu_2)}{n_1(n_1-n_2)}\right]e^{n_1 t} - \left[\frac{(n_2+\mu_1)(n_2+\mu_2)}{n_2(n_1-n_2)}\right]e^{n_2 t} \quad (11.4)$$

where

$$n_1, n_2 = \frac{-b \pm \left[b^2 - 4(\mu_1\mu_2 + \lambda_2\mu_1 + \lambda_1\mu_2)\right]^{1/2}}{2} \quad (11.5)$$

$$b = \lambda_1 + \lambda_2 + \mu_1 + \mu_2 \quad (11.6)$$

$$n_1 n_2 = \mu_1\mu_2 + \lambda_2\mu_1 + \lambda_1\mu_2 \quad (11.7)$$

$$n_1 + n_2 = \lambda_1 + \lambda_2 + \mu_1 + \mu_2 \quad (11.8)$$

$$P_1(t) = \frac{\lambda_1\mu_2}{n_1 n_2} + \left[\frac{\lambda_1 n_1 + \lambda_1\mu_2}{n_1(n_1-n_2)}\right]e^{n_1 t} - \left[\frac{(\mu_2+n_2)\lambda_1}{n_2(n_1-n_2)}\right]e^{n_2 t} \quad (11.9)$$

$$P_2(t) = \frac{\mu_1\lambda_2}{n_1 n_2} + \left[\frac{\lambda_2 n_1 + \lambda_2\mu_1}{n_1(n_1-n_2)}\right]e^{n_1 t} - \left[\frac{(\mu_1+n_2)\lambda_2}{n_2(n_1-n_2)}\right]e^{n_2 t} \quad (11.10)$$

The robot availability, $RAV(t)$, is given by

$$RAV(t) = P_0(t) \quad (11.11)$$

As time $t$ becomes very large in Equations 11.9 through 11.11, the following steady-state probability expressions are obtained:

$$P_1 = \frac{\lambda_1\mu_2}{n_1 n_2} \quad (11.12)$$

$$P_2 = \frac{\mu_1\lambda_2}{n_1 n_2} \quad (11.13)$$

$$RAV = \frac{\mu_1\mu_2}{n_1 n_2} \quad (11.14)$$

where

$P_1$ is the steady-state probability of the robot being in state 1.
$P_2$ is the steady-state probability of the robot being in state 2.
$RAV$ is the robot steady-state availability.

For $\mu_1 = \mu_2 = 0$, from Equations 11.4, 11.9, and 11.10, we obtain

$$P_0(t) = e^{-(\lambda_1+\lambda_2)t} \tag{11.15}$$

$$P_1(t) = \frac{\lambda_1}{(\lambda_1+\lambda_2)}[1 - e^{-(\lambda_1+\lambda_2)t}] \tag{11.16}$$

$$P_2(t) = \frac{\lambda_2}{(\lambda_1+\lambda_2)}[1 - e^{-(\lambda_1+\lambda_2)t}] \tag{11.17}$$

Thus, the robot reliability at time $t$ from Equation 11.15 is

$$R_r(t) = e^{-(\lambda_1+\lambda_2)t} \tag{11.18}$$

where
$R_r(t)$ is the robot reliability at time $t$.

By using the formula for mean time to failure presented in Chapter 3 and Equation 11.18, we get the following expression for the mean time to robot-related failure:

$$\begin{aligned} RMTTF &= \int_0^\infty e^{-(\lambda_1+\lambda_2)t}dt \\ &= \frac{1}{\lambda_1+\lambda_2} \end{aligned} \tag{11.19}$$

where
$RMTTF$ is the mean time to robot failure.

**EXAMPLE 11.1**

Assume that a robot can fail either due to human error or other failures (i.e., hardware and software failures) and its constant human error and other failure rates are 0.002 errors per hour and 0.007 failures per hour, respectively. The robot constant repair rate from both these failure modes is 0.09 repairs per hour.

Calculate the robot steady-state availability.

By inserting the given data values into Equation 11.14, we get

$$\begin{aligned} RAV &= \frac{(0.09)(0.09)}{(0.09)(0.09)+(0.002)(0.09)+(0.007)(0.09)} \\ &= 0.9091 \end{aligned}$$

Thus, the robot steady-state availability is 0.9091.

## 11.3 Model II

This mathematical model represents a situation where a worker performs a time-continuous robot-related task under fluctuating environments (i.e., normal and abnormal) and he/she can commit an error in either a normal or an abnormal environment. The state-space diagram for the worker carrying out a time-continuous robot-related task is shown in Figure 11.2. The numerals in the rectangles or boxes denote system states.

This mathematical model is subjected to the following assumptions:

- Errors occur independently.
- The rate of change of the environment from normal to abnormal (or stressful) or vice versa is constant.
- Worker error rates are constant.
- The worker is executing a robot-related time-continuous task.

The following symbols are associated with the diagram shown in Figure 11.2 and its associated equations:

$P_i(t)$ is the probability of the worker being in state $i$ at time $t$, for $i = 0,1,2,3$.

$\lambda_n$ is the constant error rate of the worker carrying out a robot-related task in a normal environment.

$\lambda_a$ is the constant error rate of the worker carrying out a root-related task in an abnormal or stressful environment.

$\theta_1$ is the constant transition rate from a normal environment to an abnormal or stressful environment.

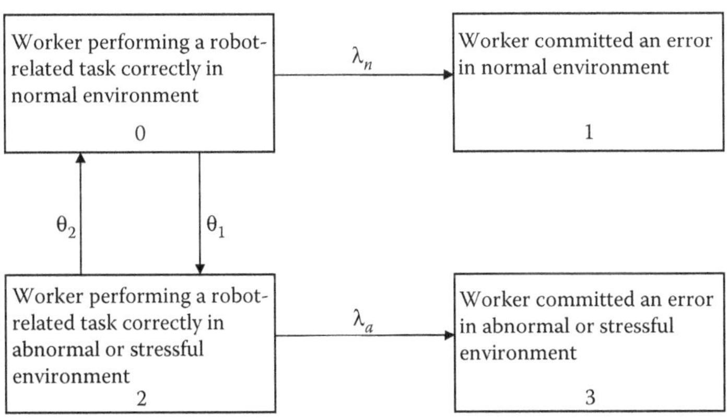


**FIGURE 11.2**
State-space diagram for the worker performing a robot-related task in fluctuating environments.

$\theta_2$ is the constant transition rate from an abnormal or stressful environment to a normal environment.

Using the Markov method, we have the following set of differential equations for the state-space diagram shown in Figure 11.2 [4]:

$$\frac{dP_0(t)}{dt} + (\lambda_n + \theta_1)P_0(t) = \theta_2 P_2(t) \tag{11.20}$$

$$\frac{dP_1(t)}{dt} = \lambda_n P_0(t) \tag{11.21}$$

$$\frac{dP_2(t)}{dt} + (\lambda_a \theta_2)P_2(t) = \theta_1 P_0(t) \tag{11.22}$$

$$\frac{dP_3(t)}{dt} = \lambda_a P_2(t) \tag{11.23}$$

At time $t = 0$, $P_0(0) = 1$ and $P_1(0) = P_2(0) = P_3(0) = 0$.

Solving Equations 11.20 through 11.23, we get

$$P_0(t) = (y_1 - y_2)^{-1}[(y_2 + \lambda_a + \theta_2)e^{y_2 t} - (y_1 + \lambda_a + \theta_2)e^{x_1 t}] \tag{11.24}$$

where

$$y_1 = \frac{-c_1 + \sqrt{c_1^2 - 4c_2}}{2} \tag{11.25}$$

$$y_2 = \frac{-c_1 - \sqrt{c_1^2 - 4c_2}}{2} \tag{11.26}$$

where

$$c_1 = \lambda_n + \lambda_a + \theta_1 + \theta_2 \tag{11.27}$$

$$c_2 = \lambda_n(\lambda_a + \theta_2) + \theta_1 \lambda_a \tag{11.28}$$

$$P_1(t) = c_4 + c_5 e^{y_2 t} - c_6 e^{y_1 t} \tag{11.29}$$

where

$$c_4 = \frac{\lambda_n(\lambda_a + \theta_2)}{y_1 y_2} \tag{11.30}$$

$$c_5 = c_3(\lambda_n + c_4 y_1) \tag{11.31}$$

$$c_3 = \frac{1}{y_2 - y_1} \tag{11.32}$$

$$c_6 = c_3(\lambda_n + c_4 y_2) \tag{11.33}$$

$$P_2(t) = \theta_1 c_3 (e^{y_2 t} - e^{y_1 t}) \tag{11.34}$$

$$P_3(t) = c_7[(1 + c_3)(y_1 e^{y_2 t} - y_2 e^{y_1 t})] \tag{11.35}$$

where

$$c_7 = \frac{\lambda_a \theta_1}{y_1 y_2} \tag{11.36}$$

The reliability of the worker in a fluctuating environment is given by

$$WR(t) = P_0(t) + P_1(t) \tag{11.37}$$

where

$WR(t)$ is the reliability of the worker in carrying out a robot-related task in a fluctuating environment at time $t$.

By using the formula for mean time to failure presented in Chapter 3 and Equation 11.37, we obtain the following expression for the mean time to error of the worker:

$$WMTTE = \int_0^\infty WR(t)dt = \frac{\lambda_a + \theta_1 + \theta_2}{c_2} \tag{11.38}$$

where

$WMTTE$ is the mean time to error of the worker.

**EXAMPLE 11.2**

Assume that a worker is carrying out a robot-related task in a fluctuating environment (i.e., normal and abnormal [stressful]) and his/her error rates are 0.005 errors per hour and 0.009 errors per hour, respectively. The constant transition rates—from a normal to an abnormal environment and vice versa—are 0.005 per hour and 0.002 per hour, respectively. Calculate the mean time to error of the worker.

By inserting the specified data values into Equation 11.38, we get

$$WMTTE = \frac{(0.009 + 0.005 + 0.002)}{(0.005)(0.009 + 0.002) + (0.005)(0.009)}$$
$$= 160 \text{ h}$$

Thus, the mean time to error of the worker is 160 h.

## 11.4 Model III

This mathematical model represents a robot system made up of a robot and a safety unit; as in industry, the inclusion of safety units/systems is often practiced because of the occurrence of robot accidents involving humans. In the model, it is assumed that after the failure of the safety unit, the robot may fail with an incident or safely and the failed safety unit is repaired.

The robot system state-space diagram is shown in Figure 11.3 and the numerals in rectangles or boxes denote system states.

The model is subjected to the following four assumptions:

- The robot system fails when the robot fails.
- Failures occur independently.
- The repaired safety unit is as good as new.
- All failure and repair rates are constant.

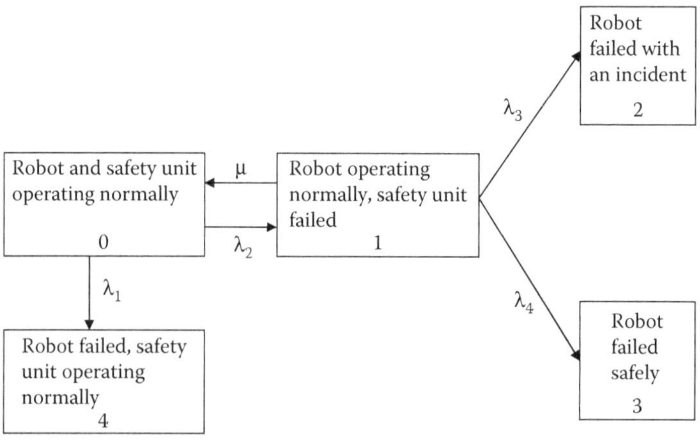


**FIGURE 11.3**
Robot system state-space diagram.

The following symbols are associated with the state-space diagram shown in Figure 11.3 and its associated equations:

- $P_j(t)$ is the probability that the robot system is in state $j$ at time $t$; for $j=0$ (robot and safety unit operating normally), $j=1$ (robot operating normally, safety unit failed), $j=2$ (robot failed with an incident), $j=3$ (robot failed safely), $j=4$ (robot failed, safety unit operating normally).
- $\mu$ is the safety unit repair rate.
- $\lambda_1$ is the robot failure rate.
- $\lambda_2$ is the safety unit failure rate.
- $\lambda_3$ is the failure rate of the robot failing with an incident.
- $\lambda_4$ is the failure rate of the robot failing safely.

Using the Markov method, we have the following set of differential equations for the state-space diagram shown in Figure 11.3 [5]:

$$\frac{dP_0(t)}{dt} + (\lambda_1 + \lambda_2)P_0(t) = \mu P_1(t) \tag{11.39}$$

$$\frac{dP_1(t)}{dt} + (\lambda_3 + \lambda_4 + \mu)P_1(t) = \lambda_2 P_0(t) \tag{11.40}$$

$$\frac{dP_2(t)}{dt} = \lambda_3 P_1(t) \tag{11.41}$$

$$\frac{dP_3(t)}{dt} = \lambda_4 P_1(t) \tag{11.42}$$

$$\frac{dP_4(t)}{dt} = \lambda_1 P_0(t) \tag{11.43}$$

At time $t=0$, $P_0(0)=1$ and $P_1(0)=P_2(0)=P_3(0)=P_4(0)=0$.

Solving Equations 11.39 through 11.43 using Laplace transforms, we obtain

$$P_0(t) = e^{-Xt} + \lambda_2\mu\left[\frac{e^{-Xt}}{(a_1+X)(a_2+X)} + \frac{e^{-a_1 t}}{(a_1+X)(a_1-a_2)} + \frac{e^{-a_2 t}}{(a_2+X)(a_2-a_1)}\right] \tag{11.44}$$

where

$$X = \lambda_2 + \lambda_1 \tag{11.45}$$

$$a_1, a_2 = \frac{-Y \pm (Y^2 - 4Z)^{1/2}}{2} \tag{11.46}$$

$$Y = X + \mu + \lambda_3 + \lambda_4 \tag{11.47}$$

$$Z = \lambda_3\lambda_2 + \lambda_4\lambda_2 + \lambda_3\lambda_1 + \lambda_4\lambda_1 + \mu\lambda_1 \tag{11.48}$$

$$P_1(t) = \lambda_2\left[\frac{e^{a_1t} - e^{a_2t}}{a_1 - a_2}\right] \tag{11.49}$$

$$P_2(t) = \frac{\lambda_3\lambda_2}{a_1a_2}\left[1 + \frac{a_1e^{a_2t} - a_2e^{a_1t}}{a_2 - a_1}\right] \tag{11.50}$$

$$P_3(t) = \frac{\lambda_4\lambda_2}{a_1a_2}\left[1 + \frac{a_1e^{a_2t} - a_2e^{a_1t}}{a_2 - a_1}\right] \tag{11.51}$$

$$P_4(t) = \frac{\lambda_1}{X}(1 - e^{-Xt}) + \lambda_2\lambda_1\mu\left[\frac{1}{a_1a_2X} - \frac{e^{-Xt}}{X(a_1 + X)(a_2 + X)} + \frac{e^{a_1t}}{a_1(a_1 + A)(a_1 - a_2)} + \frac{e^{a_2t}}{a_2(a_2 + X)(a_2 - a_1)}\right] \tag{11.52}$$

The robot system reliability is given by

$$RSR(t) = P_0(t) \tag{11.53}$$

where
$RSR(t)$ is the robot system reliability at time $t$.

By using the formula for mean time to failure presented in Chapter 3 and Equation 11.53, we get the following expression for the robot system's mean time to failure:

$$RSMTTF = \int_0^\infty RSR(t)dt = \frac{1}{X}\left[1 + \frac{\lambda_2\mu}{Z}\right] \tag{11.54}$$

where
$RSMTTF$ is the robot system's mean time to failure.

**EXAMPLE 11.3**

A robot system is made up of a robot and a safety unit and the operating robot with failed safety unit can fail either safely or with an incident. Only the failed safety unit is repaired.

Calculate the robot system's mean time to failure with the aid of Equation 11.54 for the following specified data values:

$\mu = 0.004$ repairs per hour
$\lambda_2 = 0.0001$ failures per hour
$\lambda_3 = 0.0005$ failures per hour
$\lambda_4 = 0.0003$ failures per hour
$\lambda_1 = 0.0008$ failures per hour

By substituting the above given data values into Equation 11.54, we get

$$RSMTTF = \frac{1}{X}\left[1 + \frac{(0.0001)(0.004)}{Z}\right]$$

where

$$X = 0.0001 + 0.0008$$

$$\begin{aligned} Z &= (0.0005)(0.0001) + (0.0003)(0.0001) + (0.0005)(0.0008) \\ &\quad + (0.0003)(0.0008) + (0.004)(0.0008) \\ &= \frac{1}{0.0009}\left[1 + \frac{(0.0001)(0.004)}{(0.00000392)}\right] \\ &= 1224.5 \text{ h} \end{aligned}$$

Thus, the robot system's mean time to failure is 1224.5 h.

## 11.5 Model IV

This mathematical model represents a robot system having three states: operating normally, operating unsafely, and failed. The robot system is repaired from failed and unsafe operating states. The robot system state-space diagram is shown in Figure 11.4 and the numerals in circles denote the robot system states.

The model is subjected to the following assumptions:

- All robot system failure and repair rates are constant.
- The repaired robot system is as good as new.
- Failures occur independently.

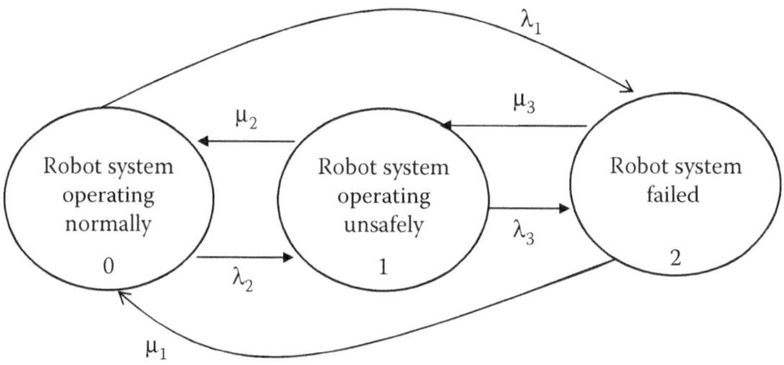


**FIGURE 11.4**
Robot system state-space diagram.

The following symbols are associated with the state-space diagram shown in Figure 11.4 and its associated equations:

$P_i(t)$ is the probability that the robot system is in state $i$ at time $t$; for $i = 0$ (robot system operating normally), $i = 1$ (robot system operating unsafely), $i = 2$ (robot system failed).

$\lambda_i$ is the robot system ith constant failure rate, where $i = 1$ means from state 0 to state 2, $i = 2$ means from state 0 to state 1, $i = 3$ means from state 1 to state 2.

$\mu_i$ is the robot system $i$th constant repair rate, where $i = 1$ means from state 2 to state 0, $i = 2$ means from state 1 to state 0, $i = 3$ means from state 2 to state 1.

Using the Markov method presented in Chapter 4, we have the following set of differential equations for the state-space diagram shown in Figure 11.4 [2,6,7]:

$$\frac{dP_0(t)}{dt} + (\lambda_1 + \lambda_2)P_0(t) = \mu_1 P_2(t) + \mu_2 P_1(t) \tag{11.55}$$

$$\frac{dP_1(t)}{dt} + (\mu_2 + \lambda_3)P_1(t) = \lambda_2 P_0(t) + \mu_3 P_2(t) \tag{11.56}$$

$$\frac{dP_2(t)}{dt} + (\mu_1 + \mu_3)P_2(t) = \lambda_1 P_0(t) + \lambda_3 P_1(t) \tag{11.57}$$

At time $t = 0$, $P_0(0) = 1$, and $P_1(0) = P_2(0) = 0$.

For a very large $t$, solving Equations 11.55 through 11.57, we get the following steady-state probability equations [7,8]:

$$P_0 = \frac{(\mu_1 + \mu_3)(\mu_2 + \lambda_3) - \lambda_3\mu_3}{D} \tag{11.58}$$

where

$$D = (\mu_1 + \mu_3)(\mu_2 + \lambda_2 + \lambda_3) + \lambda_1(\mu_2 + \lambda_3) + \lambda_1\mu_3 + \lambda_2\lambda_3 - \lambda_3\mu_3 \tag{11.59}$$

$$P_1 = \frac{\lambda_2(\mu_1 + \mu_3) + \lambda_1\mu_3}{D} \tag{11.60}$$

$$P_2 = \frac{\lambda_1\lambda_3 + \lambda_1(\mu_2 + \lambda_3)}{D} \tag{11.61}$$

where
$P_0$, $P_1$, and $P_2$ are the steady-state probabilities of the robot system being in states 0, 1, and 2, respectively.

By setting $\mu_1 = \mu_3 = 0$ in Equations 11.55 through 11.57 and solving the resulting equations, we get the following expression for the robot system reliability:

$$\begin{aligned} RSR(t) &= P_0(t) + P_1(t) \\ &= (A_1 + B_1)e^{a_1 t} + (A_2 + B_2)e^{a_2 t} \end{aligned} \tag{11.62}$$

where
$RSR(t)$ is the robot system reliability at time $t$.

$$a_1 = \frac{-M_1 + \sqrt{M_1^2 - 4L_2}}{2} \tag{11.63}$$

$$a_2 = \frac{-M_1 - \sqrt{M_1^2 - 4L_2}}{2} \tag{11.64}$$

$$M_1 = \mu_2 + \lambda_1 + \lambda_2 + \lambda_3 \tag{11.65}$$

$$M_2 = \lambda_1\mu_2 + \lambda_1\lambda_3 + \lambda_2\lambda_3 \tag{11.66}$$

$$A_1 = \frac{a_1 + \mu_2 + \lambda_3}{(a_1 - a_2)} \tag{11.67}$$

$$A_2 = \frac{a_2 + \mu_2 + \lambda_3}{(a_2 - a_1)} \tag{11.68}$$

$$B_1 = \frac{\lambda_2}{(a_1 - a_2)} \tag{11.69}$$

$$B_2 = \frac{\lambda_2}{(a_2 - a_1)} \tag{11.70}$$

By integrating Equation 11.62 over the time interval $[0, \infty]$, we get the following expression for the robot system's mean time to failure:

$$\begin{aligned} RSMTTF &= \int_0^\infty RSR(t)dt \\ &= \frac{(A_1 + B_1)}{a_1} + \frac{(A_2 + B_2)}{a_2} \end{aligned} \tag{11.71}$$

**EXAMPLE 11.4**

Assume that for a robot system, the following values of its associated parameters are given:

$\lambda_1 = 0.007$ failures per hour
$\lambda_2 = 0.004$ failures per hour
$\lambda_3 = 0.002$ failures per hour
$\mu_1 = 0.005$ repairs per hour
$\mu_2 = 0.002$ repairs per hour
$\mu_3 = 0.003$ repairs per hour

Calculate the steady-state probability of the robot system operating unsafely with the aid of Equation 11.60.

By inserting the given data values into Equation 11.60, we obtain

$$\begin{aligned} P_1 &= \frac{(0.004)(0.005 + 0.003) + (0.007)(0.003)}{D} \\ &= 0.460 \end{aligned}$$

where

$$\begin{aligned} D = {} & (0.005 + 0.003)(0.002 + 0.004 + 0.002) + (0.007)(0.002 + 0.002) \\ & + (0.007)(0.003) + (0.004)(0.002) - (0.002)(0.003) \end{aligned}$$

Thus, the steady-state probability of the robot system operating unsafely is 0.460.

## 11.6 Model V

This mathematical model represents a robot that can either fail safely or unsafely and the repair is attempted in the field; if it cannot be repaired properly, then it is taken to the repair workshop for repair [7–9]. The state-space diagram of the model is shown in Figure 11.5 and the numerals in rectangles and circles denote the robot states.

The model is subjected to the following assumptions:

- Robot failure and repair rates and the rates of taking the failed robot to the repair workshop are constant.
- Failures occur independently.
- The repaired robot is as good as new.

The following symbols are associated with the state-space diagram shown in Figure 11.5 and its associated equations:

$P_i(t)$ is the probability that the robot is in state $i$ at time $t$; for $i = 0$ (robot working normally), $i = 1$ (robot failed unsafely), $i = 2$ (robot failed safely), $i = 3$ (robot in there pair workshop).

$\lambda_1$ is the robot constant unsafe failure rate.

$\lambda_2$ is the robot constant safe failure rate.

$\lambda_3$ is the constant rate of taking the safely-failed robot to the repair workshop.

$\lambda_4$ is the constant rate of taking the unsafely-failed robot to the repair workshop.

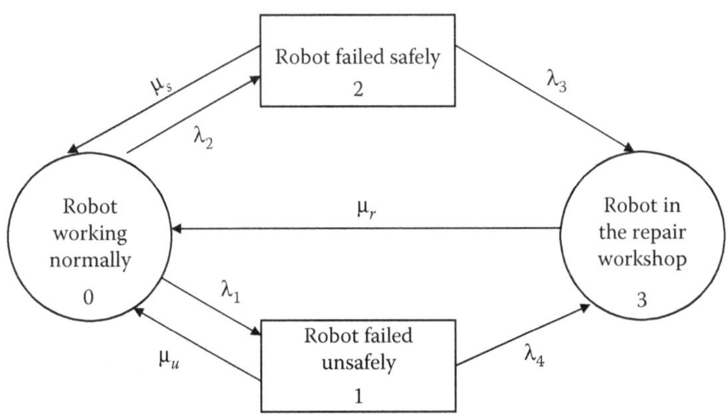


**FIGURE 11.5**
Robot state-space diagram.

$\mu_r$ is the constant repair rate of the robot from the repair workshop (i.e., state 3).

$\mu_s$ is the constant repair rate of the robot from the safely failed state (i.e., state 2).

$\mu_u$ is the constant repair rate of the robot from the unsafely failed state (i.e., state 1).

Using the Markov method presented in Chapter 4, we have the following set of differential equations for the state-space diagram shown in Figure 11.5 [2,6,7]:

$$\frac{dP_0(t)}{dt} + (\lambda_1 + \lambda_2)P_0(t) = \mu_u P_1(t) + \mu_s P_2(t) + \mu_r P_3(t) \tag{11.72}$$

$$\frac{dP_1(t)}{dt} + (\lambda_4 + \mu_u)P_1(t) = \lambda_1 P_0(t) \tag{11.73}$$

$$\frac{dP_2(t)}{dt} + (\lambda_3 + \mu_s)P_2(t) = \lambda_2 P_0(t) \tag{11.74}$$

$$\frac{dP_3(t)}{dt} + \mu_r P_3(t) = \lambda_3 P_2(t) + \lambda_4 P_1(t) \tag{11.75}$$

At time $t = 0$, $P_0(0) = 1$, and $P_1(0) = P_2(0) = P_3(0) = 0$.

By setting the derivatives equal to zero in Equations 11.72 through 11.75 and using the relationship $\Sigma_{i=0}^{3} P_i = 1$, we get the following steady-state probability equations [2,7]:

$$P_0 = \frac{A}{B} \tag{11.76}$$

and

$$P_i = m_i P_0, \quad \text{for i} = 1, 2, 3 \tag{11.77}$$

where

$$A = \mu_r(\lambda_4 + \mu_u)(\lambda_3 + \mu_s) \tag{11.78}$$

$$B = (\lambda_4 + \mu_u)\left[\mu_r(\lambda_3 + \mu_s) + \lambda_2(\mu_r + \lambda_3)\right] + \lambda_1(\lambda_3 + \mu_s)(\lambda_4 + \mu_r) \tag{11.79}$$

$$m_1 = \frac{\lambda_1}{\lambda_4 + \mu_u} \tag{11.80}$$

$$m_2 = \frac{\lambda_2}{\lambda_3 + \mu_s} \tag{11.81}$$

$$m_3 = \frac{\lambda_1\lambda_4(\lambda_3 + \mu_s) + \lambda_2\lambda_3(\lambda_4 + \mu_u)}{\lambda_2(\lambda_4 + \mu_u)(\lambda_3 + \mu_s)} \tag{11.82}$$

$P_i$ is the steady-state probability of the robot being in state $i$, for $i = 0,1,2,3$.

The robot steady-state availability and unavailability are given by

$$AV_{rs} = P_0 \tag{11.83}$$

and

$$UAV_{rs} = P_1 + P_2 + P_3 \tag{11.84}$$

where

$AV_{rs}$ is the robot's steady-state availability.

$UAV_{rs}$ is the robot's steady-state unavailability.

## 11.7 Model VI

This mathematical model is basically the same as Model V with the exception that the failed robot is not repaired. Thus, the values of $\mu_s$, $\mu_u$, and $\mu_r$ are equal to zero.

The model is subjected to the following two assumptions:

- Failures occur independently.
- Robot failure rates and the rates of taking the failed robot to the repair workshop are constant.

The symbols used in this mathematical model are same as for Model V.

Using the Markov method presented in Chapter 4, we have the following set of differential equations for the Figure 11.5 state-space diagram for $\mu_s = \mu_u = \mu_r = 0$ [2,6,7]

$$\frac{dP_0(t)}{dt} + (\lambda_1 + \lambda_2)P_0(t) = 0 \tag{11.85}$$

$$\frac{dP_1(t)}{dt} + \lambda_4 P_1(t) = \lambda_1 P_0(t) \tag{11.86}$$

$$\frac{dP_2(t)}{dt} + \lambda_3 P_2(t) = \lambda_2 P_0(t) \tag{11.87}$$

$$\frac{dP_3(t)}{dt} = \lambda_3 P_2(t) + \lambda_4 P_1(t) \tag{11.88}$$

At time $t = 0$, $P_0(0) = 1$, and $P_1(0) = P_2(0) = P_3(0) = 0$.

Solving Equations 11.85 through 11.88 using Laplace transforms, we obtain

$$P_0(t) = e^{-(\lambda_2+\lambda_1)t} \tag{11.89}$$

$$P_1(t) = X\left[e^{-(\lambda_2+\lambda_1)t} - e^{-\lambda_4 t}\right] \tag{11.90}$$

where

$$X = \frac{\lambda_1}{(\lambda_4 - \lambda_2 - \lambda_1)} \tag{11.91}$$

$$P_2(t) = Y\left[e^{-(\lambda_2+\lambda_1 t)} - e^{-\lambda_3 t}\right] \tag{11.92}$$

where

$$Y = \frac{\lambda_2}{(\lambda_3 - \lambda_2 - \lambda_1)} \tag{11.93}$$

$$P_3(t) = 1 + Xe^{-\lambda_4 t} + Ye^{-\lambda_3 t} + Ze^{-(\lambda_2+\lambda_1)t} \tag{11.94}$$

where

$$Z = -\frac{(X\lambda_4 + Y\lambda_3)}{(\lambda_2 + \lambda_1)} \tag{11.95}$$

The robot's reliability is given by

$$R_r(t) = P_0(t) = e^{-(\lambda_2+\lambda_1)t} \tag{11.96}$$

where

$R_r(t)$ is the robot reliability at time $t$.

By integrating Equation 11.96 over the time interval $[0,\infty]$, we obtain the following expression for the robot's mean time to failure:

$$\begin{aligned} RMTTF &= \int_0^{\infty} R_r(t)dt \\ &= \int_0^{\infty} e^{-(\lambda_2+\lambda_1)t}dl \\ &= \frac{1}{(\lambda_2+\lambda_1)} \end{aligned} \tag{11.97}$$

where

*RMTTF* is the robot's mean time to failure.

**EXAMPLE 11.5**

A robot can fail safely or unsafely and its constant safe and unsafe failure rates are 0.008 failures per hour and 0.002 failures per hour, respectively. Calculate the robot's mean time to failure and its reliability during 8 h mission.

By substituting the specified data values into Equations 11.97 and 11.6, we get

$$\begin{aligned} RMTTF &= \frac{1}{(0.008+0.002)} \\ &= 100 \text{ h} \end{aligned}$$

and

$$\begin{aligned} R_r(8) &= e^{-(0.008+0.002)(8)} \\ &= 0.9231 \end{aligned}$$

Thus, the robot's mean time to failure and reliability are 100 h and 0.9231, respectively.

## 11.8 Problems

1. Write an essay on mathematical models used to analyze a robot's reliability and safety.
2. Prove Equations 11.4, 11.9, and 11.10 by using Equations 11.1 through 11.3.
3. Prove Equation 11.38 by using Equation 11.37.
4. Prove that the sum of Equations 11.58, 11.60, and 11.61 is equal to unity.

5. Prove Equation 11.54 by using Equation 11.53.
6. Assume that in Figure 11.3, we have the following given values for all the five transition rates:
    a. $\lambda_1 = 0.0009$ failures per hour
    b. $\lambda_2 = 0.0002$ failures per hour
    c. $\mu = 0.003$ repairs per hour
    d. $\lambda_3 = 0.0004$ failures per hour
    e. $\lambda_4 = 0.0001$ failures per hour

    Calculate the robot system's mean time to failure.
7. Prove Equation 11.71 by using Equation 11.62.
8. Prove Equations 11.76 through 11.77 by using Equations 11.72 through 11.75.
9. Prove that the sum of Equations 11.83 through 11.84 is equal to unity.
10. Prove Equations 11.89 through 11.90, 11.92, and 11.94 with the aid of Equations 11.85 through 11.88.

## References

1. Dhillon, B.S., *Applied Reliability and Quality: Fundamentals, Methods, and Procedures*, Springer, London, 2007.
2. Dhillon, B.S., *Design Reliability: Fundamentals and Applications*, CRC Press, Boca Raton, Florida, 1999.
3. Dhillon, B.S., *Robot Reliability and Safety*, Springer-Verlag, New York, 1991.
4. Dhillon, B.S., Stochastic models for predicting human reliability, *Microelectronics and Reliability*, Vol. 24, 1984, pp. 1029–1033.
5. Dhillon, B.S., Yang, N., Reliability analysis of a repairable robot system, *Journal of Quality in Maintenance Engineering*, Vol. 2, 1996, pp. 30–37.
6. Shooman, M.L., *Probabilistic Reliability: An Engineering Approach*, McGraw-Hill Book Company, New York, 1968.
7. Dhillon, B.S., *Mine Safety: A Modern Approach*, Springer, London, 2010.
8. Dhillon, B.S., Rayapati, S.N., Reliability and availability analysis of surface transit systems, *Microelectronics and Reliability*, Vol. 24, 1984, pp. 1029–1033.
9. Dhillon, B.S., *Human Reliability: With Human Factors*, Pergamon Press, New York, 1986.

# Appendix A: Bibliography—Literature on the Reliability and Safety of Robot Systems

## A.1 Introduction

Over the years, a large number of publications on robot-system reliability, safety, and associated areas have appeared in the form of journal articles, conference proceedings articles, technical reports, and so on. This appendix presents an extensive list of selective publications related to these topics. The period covered by the listing is 1974–2013. The main objective of this listing is to provide readers with sources for obtaining additional information on robot-system reliability, safety, and associated areas.

## A.2 Publications


1. Abdallah, A., Motamed, C., Schmitt, A., Change detection for human safety in robotic environments, *Proceedings of the International Society for Optical Engineering Conference*, 1994, pp. 357–361.
2. Abdallah, A., Motamed, C., Schmitt, A., New approach for a vision-based safety device in uncontrolled robotics environments, *Proceedings of the International Society for Optical Engineering Conference*, 1995, pp. 174–181.
3. Abdi, H., Maciejewski, A.A., Nahavandi, S., Reliability maps for probabilistic guarantees of task motion for robotic manipulators, *Advanced Robotics*, Vol. 27, No. 2, 2013, pp. 81–92.
4. Abdul, S., Liu, G., Decentralized fault tolerance and fault detection of modular and reconfigurable robots with torque sensing, *Proceedings of the IEEE International Conference on Robotics and Automation*, 2008, pp. 3520–3526.
5. Abdul, S., Liu, G., Fault tolerant control of modular and reconfigurable robot with joint torque sensing, *Proceedings of the IEEE International Conference on Robotics and Biomimetics*, 2007, pp. 1236–1241.
6. Addison, J.H., Robotic safety systems and methods: Savannah River Site, *NTIS Report No. DE 8500826*, 1984; available from the National Technical Information Services (NTIS), Springfield, Virginia, USA.

7. Aghazadeh, F., Chapleski, R., Hirschfeld, R., A hazard analysis system for robotic work cells, *International Journal of Human Factors Engineering*, Vol. 8, No. 4, 1998, pp. 323–330.
8. Aghazadeh, F., Hirschfeld, R., Chapleski, R., Industrial robot use: Survey results and hazard analysis, *Proceedings of the 37th Annual Meeting of the Human Factors and Ergonomics Society*, 1993, pp. 994–998.
9. Akeel, H.A., Hardware for robotics safety systems, *Proceedings of the Interrobot West 2nd Annual Conference*, 1984, pp. 8–10.
10. Akeel, H.A., Intrinsic robot safety, *Proceedings of the Robot Safety Conference*, 1983, pp. 41–52.
11. Alayan, H., Niznik, C.A., Newcomb, R.W., Reliability of basic robot automated manufacturing networks, *Proceedings of the IEEE Southeastcon Annual Conference*, 1984, pp. 291–294.
12. Altamuro, V.M., How to achieve employee support, safety and success in your first robot installation, *Proceedings of the Robots 8 Conference*, 1984, pp. 15-1–15-8.
13. Altamuro, V.M., Working safely with iron collar worker, *National Safety News*, 1983, pp. 38–40.
14. Aluwalia, R.S., Hsu, E.Y., Sensor-based obstruction avoidance technique for a mobile robot, *Journal of Robotic Systems*, Vol. 1, 1984, pp. 331–350.
15. American Standards Institute (ANSI), *American National Standard for Industrial Robots and Robot Systems-Safety Requirements, ANSI/RIA R15.06*, New York, 1986.
16. Ancusa, V., Message redundancy in sensor networks implemented with intelligent agents, *Proceedings of the IEEE International Workshop on Robotic and Sensors Environments*, 2008, pp. 87–91.
17. Anon., Cartridge valve/brake systems increases robot safety, *Robotics World*, Vol. 1, No. 4, 1983, pp. 26–28.
18. Anon., Sensors for robot safety, *Robotics World*, Vol. 1, No. 5, 1983, pp. 16–19.
19. Anon., Protecting workers from robots, *American Machinist*, Vol. 128, No. 3, 1984, pp. 85–86.
20. Anon., Automatic robot safety shutdown system, *NASA Technical Briefs*, 1984, pp. 546–547.
21. Anon., Computing and control division colloquium on safety and reliability of complex robotic systems, *IEE Colloquium (Digest)*, No. 085, London, 1994.
22. Anon., Electrostatic liquid cleaning prevents servo-valve failures, *Robotics World*, Vol. 5, 1987, pp. 32–34.
23. Anon., HSE emphasizes a flexible approach to robot safety, *The Industrial Robot*, Vol. 14, 1987, pp. 49.
24. Anon., Robot handling provides a safe solution, *The Industrial Robot*, Vol. 12, 1985, pp. 92.
25. Anon., Robot safety: In a state of flux and jungle, *Robot News International*, 1982, pp. 3–4.
26. Anon., Safety with sophistication, *The Industrial Robot*, Vol. 11, No. 4, 1984, pp. 243–245.
27. Anon., Watchdog robot ensures safety, *Machine Design*, Vol. 55, 1983, p. 16.
28. Argote, L., Goodman, P.S., Schkade, D., The human side of robotics: How workers react to a robot, *Sloan Management Review*, 1983, pp. 33–41.
29. Asikin, D., Dolan, J., Reliability impact on planetary robotic missions, *Proceedings of the IEEE International Conference on Intelligent Robots and Systems*, 2010, pp. 4095–4100.

30. Atcitty, C.B., Robinson, D.G., Safety assessment of a robotic system handling nuclear material, *Proceedings of the ASCE Specialty Conference*, 1996, pp. 255–261.
31. Ayrulu, B., Barshan, B., Reliability measure assignment to sonar for robust target differentiation, *Pattern Recognition*, Vol. 35, No. 6, 2002, pp. 1403–1419.
32. Barabanov, V., Chirkov, A.L., Reliability of robot assisted manufacturing cells in their initial operating period, *Soviet Engineering Research*, Vol. 10, No. 1, 1990, pp. 83–85.
33. Barcheck, C., Methods for safe robotics start-up, testing, inspection and maintenance, *Proceedings of the RIA Robot Seminar*, 1985, pp. 67–73.
34. Barrett, R.J., Bell, R., Hodson, P.H., Planning for robot installation and maintenance: A safety framework, *Proceedings of the 4th British Robot Association Annual Conference*, 1981, pp. 13–22.
35. Barrett, R.J., Practical robot safety measures within a legal framework, *Proceedings of the 6th British Robot Association Annual Conference*, 1983, pp. 33–39.
36. Barrett, R.J., Robot safety and the law, in *Robot Safety*, eds. M.C. Bonney and Y.F. Yong, Springer-Verlag, Berlin, 1985, pp. 3–15.
37. Barrett, R.J., Robot safety: The HSE experience, *Proceedings of the 7th British Robot Association Annual Conference*, 1984, pp. 309–313.
38. Bary, D.J., Practical examples of safety in robot systems, *Proceedings of the International Seminar on Safety in Advanced Manufacturing*, 1987, pp. 27–36.
39. Basir, O.A., Markovian model for predicting the impact of observation conditions on the reliability of sensory systems, innovations in theory, practice and applications, *Proceedings of the IEEE International Conference on Intelligent Robots and Systems*, 1998, pp. 1646–1651.
40. Basu, A., Elnagar, A., Safety optimizing strategies for local path planning in dynamic environments, *International Journal of Robotics and Automation*, Vol. 10, No. 4, 1995, pp. 130–142.
41. Beauchamp, Y., Imbeau, D., Stobbe, T.J., Applying human factors engineering to man-robot systems, *Proceedings of the International Industrial Engineering Conference*, 1990, pp. 481–486.
42. Beauchamp, Y., Stobbe, T.J., Effects of factors on human performance in the event of an unexpected robot motion, *Journal of Safety Research*, Vol. 21, No. 3, 1990, pp. 83–96.
43. Beauchamp, Y., Stobbe, T.J., Human performance in human-robot systems, *Proceedings of the 3rd International Conference on CAD/CAM, Robotics and Factories of the Future*, 1988, pp. 33–36.
44. Becker, C. et al., Reliable navigation using landmarks, *Proceedings of the IEEE International Conference on Robotics and Automation*, 1995, pp. 401–406.
45. Ben-Lamine, M.S. et al., Mechanical impedance characteristics of robots for coexistence with humans, *Proceedings of the IEEE International Conference on Robotics and Automation*, 1997, pp. 907–912.
46. Bhatti, P.K., Rao, S.S., Reliability analysis of robot manipulators, *Proceedings of the 17th ASME Design Automation Conference*, 1987, pp. 45–53.
47. Blankenship, J., Strickland, D., Kittampton, K., Robot control for hazard prevention, *Proceedings of the American Society of Mechanical Engineers (ASME) International Computers in Engineering Conference*, 1987, pp. 164–167.
48. Bowling, A.P. et al., Reliability-based design optimization of robotic system dynamic performance, *Journal of Mechanical Design*, Vol. 129, 2007, pp. 449–454.

49. Brady, K.J. et al., Modular controller architecture for multi-arm telerobotic systems, *Proceedings of the IEEE International Conference on Robotics and Automation*, 1996, pp. 1024–1029.
50. Braman, J.M.B., Murray, R.M., Wagner, D.A., Safety verification of a fault tolerant reconfigurable autonomous goal-based robotic control system, *Proceedings of the International Conference on Intelligent Robots and Systems*, 2007, pp. 853–858.
51. Brewer, B.R., Pradhan, S., Preliminary investigation of test-retest reliability of a robotic assessment for Parkinson's disease, *Proceedings of the 32nd Annual International Conference of the IEEE EMBS*, 2010, pp. 5863–5866.
52. Broder, J., Robots seen as posing a threat to worker safety, *Los Angeles Times*, December 14, 1984.
53. Brooks, R.R., Iyengar, S.S., Robot algorithm evaluation by simulating sensor faults, *Proceedings of SPIE Conference*, Vol. 2484, 1995, pp. 394–401.
54. Brown, W.R., Ulsoy, A.G., A passive-assist design approach for improved reliability and efficiency of robots arms, *Proceedings of the IEEE International Conference on Robotics and Automation*, 2011, pp. 4927–4934.
55. Buckingham, R., Safe active robotic devices for surgery, *Proceedings of the IEEE International Conference on Systems, Man, and Cybernetics*, 1993, pp. 355–358.
56. Buckingham, R., Towards safe active robotic devices for surgery, *Industrial Robot*, Vol. 20, No. 2, 1993, pp. 8–11.
57. Burton, J., Industrial robotics: Hazards, accidents, safety applications and advanced sensor technology, *Professional Safety*, Vol. 33, 1988, pp. 28–33.
58. Caccavale, F., Walker, L.D., Observer-based fault detection for robot manipulators, *Proceedings of the IEEE International Conference on Robotics and Automation*, 1997, pp. 2881–2887.
59. Calestine, A., Park, E.H., A formal method to characterize robot reliability, *Proceedings of the Annual Reliability and Maintainability Symposium*, 1993, pp. 395–398.
60. Carlson, J., Mulphy, R.R., How UGVs physically fail in the field, *IEEE Transactions on Robotics*, Vol. 21, No. 3, 2005, pp. 423–437.
61. Carlson, J., Murphy, R.R., Nelson, A., Follow-up analysis of mobile robot failures, *Proceedings of the IEEE International Conference on Robotics and Automation*, Vol. 2004, No. 5, 2004, pp. 4987–4994.
62. Carlson, J., Murphy, R.R., Reliability analysis of mobile robots, *Proceedings of the IEEE International Conference on Robotics and Automation*, 2003, pp. 247–281.
63. Carreras, C., Walker, I.D., An interval method applied to robot reliability quantification, *Reliability Engineering and System Safety*, Vol. 70, 2000, pp. 291–303.
64. Carreras, C., Walker, I.D., Interval methods for fault-tree analysis in robotics, *IEEE Transactions on Reliability*, Vol. 50, No. 1, 2001, pp. 3–11.
65. Carreras, C., Walker, I.D., Interval methods for improved robot reliability estimation, *Proceedings of the Annual Reliability and Maintainability Symposium*, 2000, pp. 22–27.
66. Carrico, L.R., Three keys to robot safety, training, design, and implementation, *National Safety News*, 1984, pp. 81–85.
67. Cavallaro, J.R., Walker, I.D., Failure mode analysis of a proposed manipulator–based hazardous material retrieval system, *Proceedings of the ANS 7th Topical Meeting on Robotics and Remote Systems*, 1997, pp. 1096–1102.
68. Chamin, A.F., Glebov, M.R., Dytalov, V.P., Improving the reliability of robot-based sections and flexible manufacturing systems, *Chemical and Petroleum Engineering*, Vol. 24, 1988, pp. 241–245.

69. Chen, I., Effect of parallel planning on system reliability of real-time expert systems, *IEEE Transactions on Reliability*, Vol. 46, No. 1, 1997, pp. 81–87.
70. Chen, W., Hou, L., Cai, H., Research on reliability distribution method for a portable robot system, *Jiqiren/Robot*, Vol. 24, No. 1, 2002, pp. 35–38.
71. Chen, W., Hou, L., Cai, H., Two-level reliability optimizing distribution and its algorithms for series industry robot systems, *Gaojishu Tongxin/High Technology Letters*, Vol. 10, No. 2, 2000, pp. 78–81.
72. Chen, W. et al., Reliability optimizing distribution for a portable arc-welding robot system, *Harbin Gongye Daxue Xuebao/Journal of Harbin Institute of Technology*, Vol. 32, No. 1, 2000, pp. 12–14.
73. Chinellato, E. et al., Visual quality measures for characterizing planar robot grasps, *IEEE Transactions on Systems, Man, and Cybernetics*, Vol. 35, No. 1, 2005, pp. 30–41.
74. Chu, Y.X. et al., Localization algorithms performance evaluation and reliability analysis, *Proceedings of the IEEE International Conference on Robotics and Automation*, 1998, pp. 3652–3657.
75. Cohen, P.H., Chandra, M.J., Framework for the reliability modeling of robotic cells, *Proceedings of the 6th AUTOFACT Conference*, 1987, pp. 30–41.
76. Collins, J.W., Hazard prevention in automated factories, *Robotics Engineering*, Vol. 8, 1986, pp. 8–11.
77. Cop, V., Increasing the quality and reliability of industrial robots and automated workplace, in *Development in Robotics*, ed. B. Rooks, North-Holland, Amsterdam, 1983, pp. 89–98.
78. Corbel, D., Company, O., Pierrot, F., From a 3-DOF parallel redundant archi robot to a auto-calibrated archi robot, *Proceedings of the ASME International Design Engineering Technical Conference and Computers and Information in Engineering Conference*, 2007, pp. 847–855.
79. Crane, C.D. et al., Faster than real time robot simulation for plan development and robot safety, *Proceedings of the Conference on Remote Systems Technology*, 1996, pp. 14–16.
80. Crowder, R.M. et al., Maintenance of robotic systems using hypermedia and case-based reasoning, *Proceedings of the IEEE International Conference on Robotics and Automation*, 2000, pp. 2422–2427.
81. Dashiu, G. et al., Robotic assembly operation reliability, *Proceedings of the ASME Conference on Flexible Assembly Systems*, 1990, pp. 65–70.
82. Davies, B.L., Hibberd, R., Safe communication system for wheelchair-mounted medical robots, *Computing and Control Engineering Journal*, Vol. 6, No. 5, 1995, pp. 26–221.
83. Davies, B.L., Ng, W.S., Hibberd, P.D., Prostatic resection: An example safe robotic surgery, *Robotica*, Vol. 11, 1993, pp. 561–566.
84. Desai, M. et al., Effects of changing reliability on trust of robot systems, *Proceedings of the IEEE International Conference on Human-Robot Interaction*, 2012, pp. 73–80.
85. Devianayagam, S., Human factors concerns in industrial robot applications, *Proceedings of the Human Factors Society Annual Conference*, 1984, pp. 40–44.
86. Dhillon, B.S., Aleem, M.A., Report on robot reliability and safety in Canada: A survey of robot users, *Journal of Quality in Maintenance Engineering*, Vol. 6, No. 1, 2000, pp. 61–74.
87. Dhillon, B.S., Anude, O.C., Robot safety and reliability: A review, *Microelectronics and Reliability*, Vol. 33, No. 3, 1993, pp. 413–429.

88. Dhillon, B.S., Fashandi, A.R.M., Liu, K.L., Robot systems reliability and safety: A review, *Journal of Quality in Maintenance Engineering*, Vol. 8, No. 3, 2002, pp. 170–212.
89. Dhillon, B.S., Fashandi, A.R.M., Robotic systems probabilistic analysis, *Microelectronics and Reliability*, Vol. 37, No. 2, 1997, pp. 211–224.
90. Dhillon, B.S., Fashandi, A.R.M., Safety and reliability assessment techniques in robotics, *Robotica*, Vol. 15, No. 6, 1997, pp. 701–708.
91. Dhillon, B.S., Fashandi, A.R.M., Stochastic analysis of a robot machine with duplicate safety units, *Journal of Quality in Maintenance Engineering*, Vol. 5, No. 2, 1999, pp. 114–127.
92. Dhillon, B.S., Li, Z., Stochastic analysis of a maintainable robot-safety system with common-cause failures, *Journal of Quality in Maintenance Engineering*, Vol. 10, No. 2, 2004, pp. 136–147.
93. Dhillon, B.S., On robot reliability and safety: Bibliography, *Microelectronics and Reliability*, Vol. 27, 1987, pp. 105–118.
94. Dhillon, B.S., *Robot Reliability and Safety*, Springer-Verlag, New York, 1991.
95. Dhillon, B.S., Yang, N., Availability analysis of a robot with safety system, *Microelectronics and Reliability*, Vol. 36, No. 2, 1996, pp. 169–177.
96. Dhillon, B.S., Yang, N., Formulas for analyzing a redundant robot configuration with a built-in system, *Microelectronics and Reliability*, Vol. 37, No. 4, 1997, pp. 557–563.
97. Dhillon, B.S., Yang, N., Reliability analysis of a repairable robot system, *Journal of Quality in Maintenance Engineering*, Vol. 2, No. 2, 1996, pp. 30–37.
98. Dixon, W.E. et al., Fault detection for robot manipulators with parametric uncertainty: A prediction-error-based approach, *IEEE Transactions on Robotics and Automation*, Vol. 16, No. 6, 2000, pp. 689–699.
99. Dixon, W.E. et al., Fault detection for robot manipulators with parametric uncertainty: A prediction error-based approach, *Proceedings of the IEEE International Conference on Robotics and Automation*, Vol. 4, 2000, pp. 3628–3634.
100. Doming, E.E. et al., Robotic and nuclear safety for an automated/tele-operated glove box system, *Proceedings of the Conference on Remote Systems Technology*, 1992, pp. 21–25.
101. Dragone, W. et al., Failsafe brakes keep robot arms in position, *Robotics World*, Vol. 2, 1984, pp. 20–22.
102. Duggan, R.H., Jones, R.H., Khodabandehloo, K., Towards developing reliability and safety related standards using systematic methodologies, robot technology and applications, *Proceedings of the 1st Robotics Europe Conference*, 1984, pp. 90–107.
103. Eberts, R., Salvendy, G., The contributions of cognitive engineering to the safe design and operation of CAM and robotics, *Journal of Occupational Accidents*, Vol. 8, 1986, pp. 49–67.
104. Ehrenweber, R., Testing quality characteristics with a robot, *Kunststoffe Plast Europe*, Vol. 86, No. 4, 1996, pp. 12–13.
105. Etherton, J.R., Automated maintainability records and robot safety, *Proceedings of the Annual Reliability and Maintainability Symposium*, 1987, pp. 135–140.
106. Etherton, J.R., Safe maintenance guidelines for robotic workstation, Report No. 88-108, National Institute for Occupational Safety and Health (NIOSH), Morgantown, West Virginia, 1988.
107. Etherton, J.R., Systems considerations on robot and effector speed as a risk factor during robot maintenance, *Proceedings of the 8th International Safety Conference*, 1987, pp. 343–347.

108. Faber, F., Bennewitz, M., Behnke, S., Controlling the gaze direction of a humanoid robot with redundant joints, *Proceedings of the 17th IEEE International Symposium on Robot and Human Interactive Communication*, 2008, pp. 413–418.
109. Falotico, E. et al., Heat stabilization based on a feedback error learning in a humanoid robot, *Proceedings of the IEEE International Symposium on Robot and Human Interactive Communication*, 2012, pp. 449–454.
110. Fidan, B., Anderson, B.D.O., Switching control for robust autonomous robot and vehicle platoon formation maintenance, *Proceedings of the Mediterranean Conference on Control and Automation*, 2007, pp. 1–6.
111. Fluehr, P., Industrial robots and safety report, *Werkstatt and Betrieb*, Vol. 120, No. 9, 1987, pp. 733–734.
112. Folio, D., Cadenat, V., A Redundancy-based scheme to perform safe vision-based tasks amidst obstacles, *Proceedings of the IEEE International Conference on Robotics and Biometrics*, 2006, pp. 13–18.
113. Fox, D., Taking a good, hard look at robotic safety, *Robotics World*, Vol. 17, No. 1, 1999, pp. 26–29.
114. Friedrich, W.E., Robotic handling: Sensors increase reliability, *Industrial Robot*, Vol. 22, No. 4, 1995, pp. 23–26.
115. Fryman, J., New robot safety standard: New features in the design and installation of industrial robots, *Robotics World*, Vol. 18, No. 2, 2000, p. 40.
116. Gainer, C.A., Jiang, B.C., A cause and effect analysis of industrial robot accidents from four countries, *Proceedings of the Robots 11th and 17th International Symposium on Industrial Robots*, 1987, pp. 9.1–9.9.
117. Gao, X., Liao, M., Wu, X., Run-time error detection of space-robot based on adaptive redundancy, *Aircraft Engineering and Aerospace Technology: An International Journal*, Vol. 81, No. 1, 2009, pp. 14–18.
118. Garcia, M.G.A. et al., Modeling a fault-tolerant multi agent system for the control of a mobile robot using MASE methodology, *Proceedings of the 5th WSEAS International conference on Applied Computer Science*, 2006, pp. 736–744.
119. Gertman, D.I., Bruemmer, D.J., Hartley, R.S., Improving emergency response and human–robotic performance, *Proceedings of the IEEE 8th Human Factors and Power Plants and HPRCT 13th Meeting*, 2007, pp. 334–340.
120. Ghosh, K., Lemay, C., Man–machine interactions in robotics and their effect on the safety of the work place, *Proceedings of the Robot 9th Annual Conference*, 1985, pp. 19.1–19.8.
121. Ghosh, K., Safety considerations in robotics installations, *Proceedings of the 3rd Canadian CAM/CAM and Robotics Conference*, 1984, pp. 7.9–7.15.
122. Ghosh, K., Sevingy, A., Myers, M.L., Safety problems in robotics systems and some methods for their solution, *Proceedings of the 3rd International Conference on Human Factors in Manufacturing*, 1984, pp. 724–732.
123. Ghoshray, S., Yen, K.K., Comprehensive robot collision avoidance scheme by two-dimensional geometric modeling, *Proceedings of the IEEE International Conference on Robotics and Automation*, 1986, pp. 1087–1092.
124. Gopinath, P. et al., Representation and execution support for reliable robot applications, *Proceedings of the Annual Symposium on Reliability in Distributed Systems*, 1990, pp. 96–103.
125. Gordon, J.W., Curry, J.J., Reliability analysis of the RRV-1 robot, *Proceedings of the 33rd Conference on Remote Systems Technology*, 1987, pp. 210–214.

126. Graham, J.H., Fuzzy logic approach for safety and collision avoidance in robotic systems, *International Journal of Human Factors in Manufacturing*, Vol. 5, No. 4, 1995, pp. 447–457.
127. Graham, J.H., Meagher, J.F., A sensory-based robotic safety system, *IEEE Proceedings*, Vol. 132, No. 4, 1985, pp. 183–189.
128. Graham, J.H., Meagher, J.F., Derby, S.J., A safety and collision avoidance system for industrial robots, *IEEE Transactions on Industry Applications*, Vol. 22, 1986, pp. 195–203.
129. Graham, J.H., Research issues in robot safety, *Proceedings of the IEEE International Conference on Robotics and Automation*, 1988, pp. 1854–1855.
130. Granda, T. et al., Evolutionary role of humans in the human–robot system, *Proceedings of the Human Factors Society 34th Annual Meeting*, 1990, pp. 664–668.
131. Griffin, K.G., Safety requirements for industrial robotic applications, *Proceedings of the 1st International Robot Conference*, 1986, pp. 220–225.
132. Gu, Y., Zhou, H., Ma, H., Group-control on multiped robot and method of reliability, *Jiqiren/Robot*, Vol. 24, No. 2, 2002, pp. 140–144.
133. Guiochet, J., Tondu, B., Baron, C., Integration of UML in human factors analysis for safety of a medical robot for tele-echography, *Proceedings of the IEEE International Conference on Intelligent Robots and Systems*, Vol. 4, 2003, pp. 3212–3217.
134. Haass, U.L., Kuntze, H.B., Schill, W., A surveillance system for obstacle recognition and collision avoidance control in robot environment, *Proceedings of the 2nd International Conference on Robot Vision and Sensory Controls*, 1982, pp. 357–366.
135. Halberstam, E. et al., A robot supervision architecture for safe and efficient space exploration and operation, *Proceedings of the 10th Biennial International Conference on Engineering, Construction, and Operations in Challenging Environments*, 2006, pp. 1–8.
136. Hamilton, D.I., Cavallaro, J.R., Walker, I.D., Risk and fault tolerance analysis for robotics and manufacturing, *Proceedings of the 8th Mediterranean Electrotechnical Conference*, 1996, pp. 250–255.
137. Hamilton, D.L., Walker, I.D., Bennett, J.K., Fault tolerance versus performance metrics for robot systems, *Reliability Engineering & System Safety*, Vol. 53, No. 3, 1996, pp. 309–318.
138. Hamilton, J.E., Hancock, P.A., Robotic safety: Exclusion guarding for industrial operations, *Journal of Occupational Accidents*, Vol. 8, No. 1–2, 1986, pp. 69–78.
139. Harless, M., Donath, M., An intelligent safety system for unstructured human–robot interaction, *Proceedings of the Robots 9th Conference and Exposition*, 1985, pp. 117–120.
140. Harpel, B.M. et al., Analysis of robots for hazardous environments, *Proceedings of the Annual Reliability and Maintainability Symposium*, 1997, pp. 111–116.
141. Hartmann, G., Safety features illustrated in the use of industrial robots employed in production in precision mechanical/electrical industries and manufacture of appliances, *Journal of Occupational Accidents*, Vol. 8, 1986, pp. 91–98.
142. Hasegawa, Y., Industrial robot application model design for labor saving and safety promotion in press operations, *Proceedings of the 4th International Symposium on Industrial Robots*, 1974, pp. 210–213.
143. Hasegawa, Y., Sugimoto, N., Industrial safety and robots, *Proceedings of the 12th International Symposium on Industrial Robots*, 1982, pp. 9–15.

144. Haugan, K.M., Reliability in industrial robots for spray-gun applications, *Proceedings of the 2nd Conference on Industrial Robot Technology*, 1974, pp. E.7.93–E.7.98.
145. Hazon, N., Kaminka, G.A., On redundancy, efficiency, and robustness in coverage for multiple robots, *Robotics and Autonomous Systems*, Vol. 56, 2008, pp. 1102–1114.
146. Heemskerk, C., Bosman, R., HERA: A reliable and safe space robot, *Proceedings of the IEEE International Conference on Robotics and Automation*, 1993, pp. 2799–2801.
147. Helander, M., Ergonomics and safety considerations in the design of robotics workplace: A review and some priorities for research, *International Journal of Industrial Ergonomics*, Vol. 6, No. 2, 1990, pp. 127–149.
148. Helander, M., Karawan, M.H., Methods for field evaluation of safety in a robotics workplace, in *Ergonomics of Hybrid Automated System I*, ed. W. Karwowski, Elsevier, Amsterdam, 1988, pp. 403–410.
149. Helander, M., Karawn, M.H., Etherton, J., A model of human reaction time to dangerous robot arm movements, *Proceedings of the Human Factors Society 31st Annual Meeting*, 1987, pp. 191–195.
150. Henkel, S., Robots and safety: An industry overview, *Robotics Age, The Journal of Intelligent Machine*, Vol. 7, 1985, pp. 26–28.
151. Henstridge, F., Efficiency and safety found in robotic surveying, *Public Works*, Vol. 125, No. 2, 1994, pp. 26–28.
152. Hirschfeld, RA., Aghazadeh, F., Chapleski, R.C., Survey of robot safety in industry, *International Journal of Human Factors in Manufacturing*, Vol. 3, No. 4, 1993, pp. 369–379.
153. Hishino, S., Ota, J., Design of an automated transportation system in a seaport container terminal for the reliability of operating robots, *Proceedings of the International Conference on Intelligent Robots and Systems*, 2007, pp. 4259–4264.
154. Hosaka, S., Shimizu, Y., Hayashi, T., Development of a functional fail-safe control for advanced robots, *Advanced Robotics*, Vol. 8, No. 5, 1994, pp. 477–495.
155. Hoshino, S., Ota, J., Reactive robot control with hybrid operational models in a seaport container terminal considering system reliability, *Proceedings of the IEEE/RSJ International Conference on Intelligent Robots and Systems*, 2008, pp. 143–148.
156. Hudson, M.P., Pull, D.V., Intangible light barriers for the safeguarding of industrial robot cells and other automated high risk work areas, *Proceedings of the 8th British Robot Association Annual Conference*, 1985, pp. 177–179.
157. Hulfachor, R., Safety considerations and robotic welding, *Robotics Today*, 1987, pp. 24–26.
158. Husband, T.M., Managing robot maintenance, *Proceedings of the 6th British Robot Association Annual Conference*, 1983, pp. 53–60.
159. Ikuta, K., Nokata, M., General evaluation method of safety for human-care robots, *Proceedings of the IEEE International Conference on Robotics and Automation*, 1999, pp. 2065–2072.
160. Inagaki, S., Sato, K., Summary report on the status of safety engineering for industrial robots in the United States, *Katakana/Robot*, Vol. 114, 1997, pp. 17–21.
161. Isozaki, Y., Ohnishi, K., Nagashima, T., New generation maintenance and inspection robots for nuclear power plant, *Proceedings of the International Conference on Offshore Mechanics and Arctic Engineering*, 1997, pp. 77–84.

162. Jacob, P., Surface ESD (ESDFOS) in assembly fab machineries as a functional and reliability risk-failure analysis, tool diagnosis and on-site-remedies, *Microelectronics Reliability*, Vol. 48, 2008, pp. 1608–1612.
163. Jiang, B.C. et al., Evaluation machine guarding technique for robot guarding, *Robotics and Autonomous Systems*, Vol. 7, No. 4, 1991, pp. 299–308.
164. Jiang, B.C., Gainer, C.A., A cause and effect analysis of robot accidents, *Journal of Occupational Accidents*, Vol. 9, 1987, pp. 27–45.
165. Jiang, B.C., Otto, S.H., Procedure analysis for robot system safety, *International Journal of Industrial Ergonomics*, Vol. 6, No. 2, 1990, pp. 96–117.
166. Jin, Q., Sugasawa, Y., Seya, K., Probabilistic behavior and reliability analysis for a multi-robot system by applying Petri Net and Markov renewal process theory, *Microelectronics and Reliability*, Vol. 29, 1989, pp. 993–1001.
167. Jones, R.H., Dawson, S.J., People and robots: Their safety and reliability, *Proceedings of the 7th British Robot Association Conference*, 1983, pp. 156–158.
168. Jones, R.H., Dawson, S.J., Strategies for ensuring safety with industrial robot systems, *Omega International Journal of Management Science*, Vol. 14, 1986, pp. 287–297.
169. Jones, R.H., Dawson, S.J., The role of hardware, software and people in safeguarding robot production systems, *Proceedings of the 15th International Symposium on Industrial Robots*, 1985, pp. 557–568.
170. Junyao, G. et al., Fault-tolerant and high reliability space robot design and research, *Proceedings of the International Joint Conference on Neural Networks*, 2008, pp. 2413–2417.
171. Kabayashi, F., Arai, F., Fakuda, T., Sensor selection by reliability based on possibility measure, *Proceedings of the IEEE International Conference on Robotics and Automation*, 1999, pp. 2614–2619.
172. Kanatani, K., Ohta, N., Optimal robot self-localization and reliability evaluation, *Lecture Notes in Computer Science*, Vol. 1407, 1998, pp. 796–800.
173. Kanda, T., Ishiguro, H., An approach for a social robot to understand human relationships: Friendship estimation through interaction with robots, *Interaction Studies*, Vol. 7. No. 3, 2006, pp. 369–403.
174. Kanda, T., Ishiguro, H., Friendship estimation model for social robots to understand human relationships, *Proceedings of the IEEE International Workshop on Robot and Human Interactive Communication*, 2004, pp. 539–544.
175. Kannan, B., Parkar, L.E., Metrics for quantifying system performance in intelligent, fault-tolerant multi-robot teams, *Proceedings of the IEEE International Conference on Intelligent Robots and Systems*, 2007, pp. 951–958.
176. Karwowski, W. et al., Estimation of safe distance from the robot arm as a guide for limiting slow speed of robot motions, *Proceedings of the Human Factors Society Annual Conference*, 1993, pp. 992–996.
177. Karwowski, W. et al., Human perception of the work envelope of an industrial robot, *Journal of Occupational Accidents*, Vol. 10, 1988, pp. 116–120.
178. Karwowski, W. et al., Perception of safety zone around industrial robot, *Journal of Occupational Accidents*, Vol. 10, 1988, pp. 230–235.
170. Karwowski, W., Rahimi, M., Worker selection of safe speed and idle condition in simulated monitoring of two industrial robots, *Ergonomics*, Vol. 34, No. 5, 1991, pp. 531–546.
180. Kehoe, E.J., Practical robot safety, *Robotics Today*, 1985, pp. 38–41.

181. Kemeny, Z., Design and evaluation environment for collision-free motion planning of cooperating redundant robots, *Periodica Polytechnica Ser. El, Eng.*, Vol. 43, No. 3, 1999, pp. 189–198.
182. Khodabandehloo, K., Analyses of robot systems using fault and event trees: Case studies, *Reliability Engineering & System Safety*, Vol. 53, No. 3, 1996, pp. 247–264.
183. Khodabandehloo, K., Duggan, F., Husband, T.M., Reliability assessment of industrial robots, *Proceedings of the 14th International Symposium on Industrial Robots*, 1984, pp. 203–220.
184. Khodabandehloo, K., Duggan, F., Husband, T.M., Reliability of industrial robots: A safety viewpoint, *Proceedings of the 7th British Robot Association Conference*, 1984, pp. 143–147.
185. Khodabandehloo, K., Robot safety and reliability, in *Human-Robot Interaction*, eds. M. Rahimi and W. Karwowski, Taylor & Francis, London, 1992, pp. 121–160.
186. Kim, S., Hamel, W.R., Fault detection of tool/load grasping for telerobotics using neural networks, *Proceedings of the International Conference on Advanced Robotics*, 2005, pp. 864–869.
187. Kim., S. et al., Error analysis and calibration of camera/LRF sensor head for telerobotic systems, *Proceedings of the International Conference on Intelligent Robots and Systems*, 2006, pp. 5195–5200.
188. Kinsley, J., Sophisticated sensor system enhances robot safety, *Electrical Construction and Maintenance*, Vol. 83, 1984, pp. 53–54.
189. Kittiampton, K., Sneckenberger, J., A safety control system for a robotic work station, *Proceedings of the American Control Conference*, 1985, pp. 90–93.
190. Kobayashi, F., Arai, F., Fukuda, T., Sensor selection by reliability based on possibility measure, *Proceedings of the IEEE International Conference on Robotics and Automation*, Vol. 4, 1999, pp. 2614–2619.
191. Kochekali, H. et al., Factor affecting robot performance, *Industrial Robot*, Vol. 18, No. 1, 1991, pp. 9–13.
192. Koenig, S., Simmons, R.G., Unsupervised learning of probabilistic models for robot navigation, *Proceedings of the IEEE International Conference on Robotics and Automation*, 1996, pp. 2301–2308.
193. Koker, R., Reliability-based approach to the inverse kinematics solution of robots using Elman's networks, *Engineering Applications of Artificial Intelligence*, Vol. 18, No. 6, 2005, pp. 685–693.
194. Korayem, M.H., Iravani, A., Improvement of 3P and 6R mechanical robots reliability and quality applying FMEA and QFD approaches, *Robotics and Computer-Integrated Manufacturing*, Vol. 24, 2008, pp. 472–487.
195. Kotake, S., Safety education for people working with industrial robots, *Advanced Robotics*, Vol. 3, No. 1, 1989, pp. 75–80.
196. Kuivanen, R., Experiences from the use of an intelligent safety sensor with industrial robots, in *Ergonomics of Hybrid Automated System I*, ed. W. Karwowski, Elsevier, Amsterdam, 1988, pp. 553–559.
197. Kuntze, H.B., Schill, W., Methods for collision avoidance in computer controlled industrial robots, *Proceedings of the 12th International Symposium on Industrial Robots*, 1982, pp. 519–530.
198. Lankenau, A., Meyer, O., Krieg-Brueckner, B., Safety in robotics: The Bremen autonomous wheelchair, *Proceedings of the International Workshop on Advanced Motion Control*, 1998, pp. 524–529.

199. Lauridsen, K., Reliability of remote manipulator systems for use in radiation environments, *Proceedings of the Computing and Control Division Colloquium on Safety and Reliability of Complex Robotic Systems*, 1994, pp. 1.1–1.5.
200. Lee, C., Xu, Y., Online, interactive learning of gestures for human/robot interfaces, *Proceedings of the IEEE International Conference on Robotics and Automation*, 1996, pp. 2982–2987.
201. Lee, S. et al., Human and robot integrated teleoperation, *Proceedings of the IEEE International Conference on Systems, Man, and Cybernetics*, Vol. 2, 1998, pp. 1213–1218.
202. Lendvay, M., Accelerating reliability test of electromechanical contacts to robot controlling, *Proceedings of the International Conference on Intelligent Engineering Systems*, 1997, pp. 421–425.
203. Lester, W.A., Lannon, R.P., Bellandi, R., Recommendations for maintenance of robots, Part I, *Industrial Engineering*, Vol. 17, 1985, pp. 28–30.
204. Lester, W.A., Lannon, R.P., Bellandi, R., Recommendations for maintenance of robots, Part II, *Industrial Engineering*, Vol. 17, 1985, pp. 30–35.
205. Letchmanan, R. et al., Fault evaluation of relative-coupled BLDC drives for multi-facet mobile robot with distributed speed factors, *Proceedings of the IEEE Conference on Vehicle Power and Propulsion*, 2006, pp. 1–6.
206. Leuschen, M.L., Walker, I.D., Cavallaro, J.R., Evaluating the reliability of prototype degradable systems, *Reliability Engineering and System Safety*, Vol. 72, No. 1, 2001, pp. 9–20.
207. Leuschen, M.L., Walker, I.D., Cavallaro, J.R., Robot reliability through fuzzy Markov models, *Proceedings of the Annual Reliability and Maintainability Symposium*, 1998, pp. 209–214.
208. Lewis, C.L., Maciejewski, A.A., Example of failure tolerant operation of a kinematically redundant manipulator, *Proceedings of the IEEE International Conference on Robotics and Automation*, 1994, pp. 1380–1387.
209. Lim, H., Tanie, K., Human safety mechanisms of human-friendly robots: Passive viscoelastic trunk and passively movable base, *International Journal of Robotics Research*, Vol. 19, No. 4, 2000, pp. 307–335.
210. Lin, C., Wang, M.J., Hybrid fault tree analysis using fuzzy sets, *Reliability Engineering & System Safety*, Vol. 58, No. 3, 1997, pp. 205–213.
211. Linger, M., How to design safety systems for human protection in robot applications, *Proceedings of the 14th International Symposium on Industrial Robots*, 1984, pp. 119–129.
212. Linger, M., Sjostorm, H., Palmers, G., How to design safety functions in the control system and for the grippers of industrial robot, *Proceedings of the 15th International Symposium on Industrial Robots*, 1985, pp. 569–577.
213. Liu, B. et al., Detection of range errors due to occlusion in separated transceiver LADARS, *Proceedings of the ICARCV Conference on Control, Automation, Robotics, and Vision*, 2004, pp. 443–448.
214. Liu, T.S., Wang, J.D., Reliability approach to evaluating robot accuracy performance, *Mechanism & Machine Theory*, Vol. 29, No. 1, 1994, pp. 83–94.
215. Luck, R., Ray, A., Redundancy management and failure detection of ultrasonic ranging sensors for robotic applications, *Proceedings of the USA–Japan Symposium on Flexible Automation*, 1988, pp. 557–564.
216. Lueth, T., Extensive manipulation capabilities and reliable behavior at autonomous robot assembly, *Proceedings of the IEEE International Conference on Robotics and Automation*, 1994, pp. 3495–3500.

217. Lueth, T.C., Nassal, U.M., Rembold, U., Reliability and integrated capabilities of locomotion and manipulation for autonomous robot assembly, *Robotics and Autonomous Systems*, Vol. 14, No. 2–3, 1995, pp. 185–198.
218. Lueth, T.C., Rembold, U., Extensive manipulation capabilities and reliable behavior at autonomous robot assembly, *Proceedings of the IEEE International Conference on Robotics and Automation*, 1994, pp. 3495–3500.
219. Lumelsky, V., Cheung, E., Towards safe real-time robot teleoperation: Automatic whole sensitive arm collision avoidance frees the operator for global control, *Proceedings of the IEEE International Conference on Robotics and Automation*, 1991, pp. 797–802.
220. Luridsen, K., Christensen, P., Kongso, H.E., Assessment of the reliability of robotic system for use in radiation environments, *Reliability Engineering and System Safety*, Vol. 53, 1996, pp. 265–276.
221. Luyi, C., Xiaoqing, G., Zhenyuan, X., Reliability forecasting for industrial robot based on genetic algorithm and RBF neural network, *Proceedings of the 2nd International Conference on Mechanical and Electronics Engineering*, 2010, pp. V2-362–V2-365.
222. Lynne, E.P., Alliance: An architecture for fault tolerant multi-robot operation, *IEEE Transactions on Robotics and Automation*, Vol. 14, No. 2, 1998, pp. 220–240.
223. Maier, T., Volta, G., Wilikens, M., Reliability of robotics: An overview with identification of specific aspects related to remote handling in fusion machines, *Fusion Engineering and Design*, Vol. 29, Part C, 1995, pp. 286–297.
224. Marton, T., Pulasi, J.L., Assessment and development of HF related safety design for industrial robots, *Proceedings of the Human Factors Society 31st Annual Meeting*, 1987, pp. 176–180.
225. Matsuura, D., Iwatsuki, N., Okada, M., Redundancy optimization of hyper redundant robots based on movability and assistability, *Proceedings of the International Conference on Intelligent Robots and Systems*, 2007, pp. 2534–2539.
226. McAlinden, J.J., Using robotics as occupational health and safety control strategy, *The Industrial Robot*, Vol. 22, No. 1, 1995, pp. 14–17.
227. McCulloch, W.H., Safety analysis requirements for robotic systems in DOE nuclear facilities, *Proceedings of the ASCE Specialty Conference*, 1996, pp. 235–240.
228. Mcinroy, J.E., Saridis, G.N., Entropy searches for robotic reliability assessment, *Proceedings of the IEEE International Conference on Robotics and Automation*, 1993, pp. 935–940.
229. McInroy, J.E., Saridis, G.N., Reliable automatic plan selection for visual robotic positioning, *Proceedings of the 30th Conference on Decision and Control*, 1991, pp. 1567–1572.
230. McKinnon, R., Robots: Are they automatically safe?, *Protection*, Vol. 17, No. 5, 1980, pp. 5–6.
231. Meagher, J., Derby, S., Graliam, J., Robot safety/collision avoidance, *Professional Safety*, Vol. 28, No. 12, 1983, pp. 14–16.
232. Menon, C., Murphy, M., Sitti, M., Gecko inspired surface climbing robots, *Proceedings of the IEEE Conference on Robotics and Automation*, 2004, pp. 431–436.
233. Michaelson, D.G., Jiang, J., Modelling of redundancy in multiple mobile robots, *Proceedings of the American Control Conference*, 2000, pp. 1083–1087.
234. Mihalasky, J., The impact of robots on product reliability, *Proceedings of the Annual Reliability and Maintainability Symposium*, 1985, pp. 464–466.

235. Mladnov, M., Mock, M., Grosspietsch, K.E., Fault monitoring and correction in a walking robot using LMS filters, *Proceedings of the International Workshop on Intelligent Solutions in Embedded Systems*, 2008, pp. 1–10.
236. Monteverde, V., Tosunoglu, S., Development and application of a fault tolerance measure for serial and parallel robotic structures, *International Journal of Modeling and Simulation*, Vol. 19, No. 1, 1999, pp. 45–51.
237. Monteverde, V., Tosunoglu, S., Effect of kinematic structure and dual actuation on fault tolerance of robot manipulators, *Proceedings of the IEEE International Conference on Robotics and Automation*, 1997, pp. 2902–2907.
238. Moon, I., Joung, S. Kum, Y., Safe and reliable intelligent wheelchair robot with human-robot interaction, *Proceedings of the IEEE International Conference on Robotics and Automation*, Vol. 4, 2002, pp. 3595–3600.
239. Moorign, B., Pack, T., Aspect of robot reliability, *Robotica*, Vol. 5, Part 3, 1987, pp. 232–230.
240. Morales, A. et al., Learning to predict grasp reliability for a multifinger robot hand by using visual features, *Proceedings of the Eighth IASTED International Conference on Artificial Intelligence and Soft Computing*, 2004, pp. 249–254.
241. Morales, A. et al., An active learning approach for assessing robot grasp reliability, *Proceedings of the IEEE/RSJ International Conference on Intelligent Robots and Systems (IROS)*, Vol. 1, 2004, pp. 485–490.
242. Morita, T., Sugano, S., A technology map for standardizing safety measurements of human symbiotic robots, *Advanced Robotics*, Vol. 13, No. 3, 1999, pp. 307–308.
243. Morita, T., Sugano, S., Double safety measure for human symbiotic manipulator, *Proceedings of the IEEE/ASME International Conference on Advanced Intelligent Mechatronics*, 1997, pp. 130–131.
244. Motamed, C., Schmitt, A., Vision based safety device for uncontrolled robotic environments, *Proceedings of the IEEE International Conference on Systems, Man, and Cybernetics*, 1993, pp. 528–533.
245. Muldau, H.H.V., Safety at work-place using industrial robots, *Proceedings of the 8th International Symposium on Industrial Robots*, 1978, pp. 210–215.
246. Murakami, M., A safety failover subsystem for intelligent mobile robots, *Proceedings of the International Conference on Control, Automation and Systems*, 2007, pp. 2493–2498.
247. Murakami, M., Development of a duplex computer system for humanoid robot applications: Design of the safety failover subsystem, *Proceedings of the 33rd Annual Conference of the IEEE Industrial Electronics Society*, 2007, pp. 2783–2788.
248. Murakami, M., Fault tolerance design for computers used in humanoid robots, *Proceedings of the International Conference on Intelligent Robots and Systems*, 2007, pp. 2301–2307.
249. Musto, J.C., Saridis, G.N., Entropy-based reliability analysis for intelligent machines, *IEEE Transactions on Systems, Man, and Cybernetics, Part B: Cybernetics*, Vol. 27, No. 2, 1997, pp. 239–244.
250. Myers, J.K., Agin, G.J., A supervisory collision avoidance system for robot controllers, *Proceedings of the Annual Meeting of American Society of Mechanical Engineers*, 1982, pp. 59–63.
251. Myklebust, E., Thiel, E., Practical robot safety systems, *Proceedings of the 16th International Symposium on Industrial Robots*, 1986, pp. 1137–1146.

252. Nagamachi, M., Anayama, Y., An ergonomic study of the industrial robot, Part I: The experiments of unsafe behavior on robot manipulation, *Japanese Journal of Ergonomics*, Vol. 19, 1983, pp. 259–264.
253. Nagamachi, M. et al., A human factors study of the industrial robot, Part II: Human reliability on robot manipulation, *Japanese Journal of Ergonomics*, Vol. 20, 1984, pp. 55–64.
254. Nagamachi, M., Two fatal accidents due to robots in Japan, in *Ergonomics of Hybrid Automated System I*, ed. W. Karwowski, Elsevier, Amsterdam, 1988, pp. 391–396.
255. Narita, M., Shimamura, M., Oya, M., Reliable protocol for robot communication on web services, *Proceedings of the International Conference on Cyberworlds*, 2005, pp. 210–220.
256. Narita, S., Ohkami, Y., Development of distributed controller software for improving robot performance and reliability, *Proceedings of the IEEE/RSJ International Conference on Intelligent Robots and Systems (IROS)*, Vol. 3, 2004, pp. 2384–2389.
257. Navarro-Alarcon, D., Parra-Vega, V., Olguin-Diaz, E., Minimum set of feedback sensors for high performance decentralized cooperative force of redundant manipulators, *Proceedings of the IEEE International Workshop on Robotic and Sensors Environments*, 2008, pp. 114–119.
258. Newkirk, J.T., Bowling, A.P., Renaud, J.E., Workspace characterization of a robotic system using reliability-based design optimization, *Proceedings of the IEEE International Conference on Robotics and Automation*, 2008, pp. 3958–3963.
259. Ng., W.S., Tan, C.K., On safety enhancements for medical robots, *Reliability Engineering and System Safety*, Vol. 54, 1996, pp. 35–45.
260. Nicolaisen, P., The development of safety devices to fit the problems-example: Industrial robots, *Industrie Robotor*, Vol. 102, No. 73, 1980, pp. 164–169.
261. Noro, K., Okada, Y., Robotization and human factors, *Ergonomics*, Vol. 26, 1983, pp. 985–1000.
262. Ntuen, C.A., Park, E.H., Formal method to characterize robot reliability, *Proceedings of the Annual Reliability and Maintainability Symposium*, 1993, pp. 395–397.
263. Okina, S. et al., Study of a self-diagnosis system for an autonomous mobile robot, *Advanced Robotics*, Vol. 14, No. 5, 2000, pp. 339–341.
264. Olex, M.B., Shulman, H.G., Human factors efforts in robotics system design, *Proceedings of the 13th International Symposium on Industrial Robots*, 1983, pp. 9.29–9.36.
265. Otsuka, K., Inspection robot system of electronic unit manufacturing, *Robot*, No. 111, 1996, pp. 22–27.
266. Parker, L.E., Draper, J.V., Robotics applications in maintenance and repair, in *Handbook of Industrial Robotics*, ed. S. Y. Nof, John Wiley and Sons, New York, 1999, pp. 1023–1036.
267. Parsons, M.H., Robot system safety issues best considered in design phase, *Occupational Health and Safety*, Vol. 53, No. 8, 1984, pp. 38–42.
268. Patooghy, A. et al., A solution to single point of failure using voter replication and disagreement detection, *Proceedings of the 2nd IEEE International Symposium on Dependable, Automatic and Secure Computing*, 2006, pp. 171–176.
269. Pegman, G., Reed, J., Safety and usability of telerobotic systems, *Proceedings of the International Conference on Nuclear Engineering*, 1997, pp. 1935–1941.

270. Percival, N., Safety aspects of robots and flexible manufacturing systems, *Proceedings of the 1st International Conference on Human Factors in Manufacturing*, 1984, pp. 179–183.
271. Piggin, R., A new approach to robotic safety: Safety BUS p at BMW, *Industrial Robot*, Vol. 29, No. 6, 2002, pp. 524–529.
272. Pirjanian, P., Reliable reaction, *Proceedings of the IEEE/SICE/RSJ International Conference on Multisensor Fusion and Integration for Intelligent Systems*, 1996, pp. 158–165.
273. Polakoff, P.L., Man's marriage to robotics: A "For Better or Worse" Union, *Occupational Health and Safety*, Vol. 54, No. 4, 1985, pp. 24–25.
274. Pollard, B.W., RAM for robots: Reliability, availability, maintainability, *Robotics Today*, 1980, pp. 51–53.
275. Pomerleau, D.A., Reliability estimation for neural network based autonomous driving, *Robotics and Autonomous Systems*, Vol. 12, No. 3–4, 1994, pp. 113–119.
276. Potter, R.D., Safety for robotics, *Proceedings of the American Society of Safety Engineers Professional Development Conference*, 1983, pp. 171–183.
277. Pouliezos, A., Stavrakakis, G.S., Fast fault diagnosis for industrial processes applied to the reliable operation of oobotic systems, *International Journal of Systems Science*, Vol. 20, 1989, pp. 1233–1257.
278. Pruski, A., Surface contact sensor for robot safety, *Sensor Review*, Vol. 6, No. 3, 1986, pp. 143–144.
279. Rachkov, M., Safety systems of technological climbing robots, *Proceedings of the IEEE International Symposium on Industrial Electronics*, 1997, pp. 660–665.
280. Rahimi, M., Karwowski, W., Research paradigm in human-robot interaction, *International Journal of Industrial Ergonomics*, Vol. 5, No. 1, 1990, pp. 59–71.
281. Rahimi, M., System safety approach to robot safety, *Proceedings of the 28th Human Factors Society Annual Meeting*, 1984, pp. 102–106.
282. Rahimi, M., System safety for robots: Ali energy barrier analysis, *Journal of Occupational Accidents*, Vol. 8, 1986, pp. 127–138.
283. Rahimi, M., Xiadong, X., Framework for software safety verification of industrial robot operations, *Computers and Industrial Engineering*, Vol. 20, No. 2, 1991, pp. 279–287.
284. Ramachandran, S., Nagarajan, T., Sivaprasad, N., Reliability studies on assembly robots using the finite element method, *Advanced Robotics*, Vol. 7, No. 4, 1993, pp. 385–393.
285. Ramachandran, V., Vajpayee, S., Safety in robotic installations, *Robotics, and Computer Integrated Manufacturing*, Vol. 3, No. 3, 1987, pp. 301–308.
286. Ramirez, C.A., Artificial intelligence applied to robot fail safe operations, *Proceedings of the Robot 9 Conference on Current Issues and Future Concerns*, 1985, pp. 9.21–9.37.
287. Ramirez, C.A., Safety of robot, in *Handbook of Industrial Robots*, ed. S.Y. Nof, John Wiley and Sons, New York, 1985, pp. 131–148.
288. Rao, S.S., Bhatti, P.K., Probabilistic approach to manipulator kinematics and dynamics, *Reliability Engineering and System Safety*, Vol. 72, 2001, pp. 47–58.
289. Reinkensmeyer, D.J., Wolbrecht, E., Bobrow, J., A computational model of human-robot load sharing during robot-assisted arm movement training after stroke, *Proceedings of the 29th Annual International Conference of the IEEE Engineering in Medicine and Biology Society*, 2007, pp. 4019–4023.

290. Renaud, P., Cervera, E., Martinet, P., Towards a reliable vision-based mobile robot formation control, *Proceedings of the IEEE/RSJ International Conference on Intelligent Robots and Systems*, Vol. 4, 2004, pp. 3176–3181.
291. Roston, G.P., Dowling, K., Drivetrain design, incorporating redundancy, for an autonomous walking robot, *Proceedings of the ASCE Specialty Conference on Robotics for Challenging Environments Conference*, 1994, pp. 184–192.
292. Rovetta, A., Telerobotic surgery control and safety, *Proceedings of the IEEE International Conference on Robotics and Automation*, 2000, pp. 2895–2900.
293. Rudall, R.H., Automation and robotics world-wide reports and surveys, *Robotica*, Vol. 14, 1996, pp. 243–151.
294. Russel, J., Robot safety considerations: A check list, *Professional Safety*, Vol. 28, 1983, pp. 36–37.
295. Russell, R.A. et al., A robotic system to locate hazardous chemical leaks, *Proceedings of the IEEE International Conference on Robotics and Automation*, 1995, pp. 556–561.
296. Rybski, P. et al., Sensor fusion for human safety in industrial work-cells, *Proceedings of the IEEE/RSJ International Conference on Intelligent Robots and Systems*, 2012, pp. 1–8.
297. Sakai, H., Amasaka, K., The robot reliability design and improvement method and the advanced Toyota production system, *Industrial Robot*, Vol. 34, No. 4, 2007, pp. 310–316.
298. Salini, D., Reliability prediction with Mil handbook and field data collection in FMS, and robotics, *Microelectronics and Reliability*, Vol. 29, No. 3, 1989, pp. 415–418.
299. Salvendy, G., Review and appraisal of human aspects in planning robotic systems, *Behavior and Information Technology*, Vol. 2, 1983, pp. 262–287.
300. Sanderson, A.C., Modular robotics: Design and example, *Proceedings of the IEEE Conference on Emerging Technologies and Factory Automation*, Vol. 2, 1996, pp. 460–466.
301. Sanderson, L.M., Collins, J.W., McGlothlin, J.D., Robot-related fatality involving a US manufacturing plant employee: Case report and recommendations, *Journal of Occupational Accidents*, Vol. 8, 1986, pp. 13–23.
302. Sato, Y., Inone, K., Kumamoto, H., The safety assessment of the human–robot systems, *Bulletin of JSME*, Vol. 29, No. 250, 1980, pp. 1356–1361.
303. Sato, Y., Inone, K., Kumamoto, H., The safety assessment of the human–robot systems, *Bulletin of JSME*, Vol. 29, No. 256, 1986, pp. 3618–3525.
304. Sato, Y., Inone, K., Kumamoto, H., The safety assessment of the human–robot systems, *Bulletin of JSME*, Vol. 29, No. 257, 1986, pp. 3645–3651.
305. Sato, Y., Inoue, K., Heneley, E.J., Action chain model for the design of the hazard control systems for robots, *IEEE Transactions on Reliability*, Vol. 39, No. 2, 1990, pp. 151–157.
306. Sato, Y., Inoue, K., Heneley, E.J., Hazard assessment of industrial robots, *Proceedings of the USA-Japan Symposium on Flexible Automation*, 1988, pp. 703–708.
307. Savsar, M., Reliability analysis of a flexible manufacturing cell, *Reliability Engineering & System Safety*, Vol. 67, No. 2, 2000, pp. 147–152.
308. Schneider, D.L., Tesar, D., Barnes, J.W., Development and testing of a reliability performance index for modular robotic systems, *Proceedings of the Annual Reliability and Maintainability Symposium*, 1994, pp. 263–271.

309. Schreiber, R.R., Robot safety: A shared responsibility, *Robotics Today*, 1983, pp. 61–65.
310. Seim, B., Beutler, B., Stringent safety guidelines bring safeguarding devices into focus, *Robotics World*, Vol. 11, No. 3, 1993, pp. 16–18.
311. Seward, D., Margrave, F., LUCIE the robot excavator-design for system safety, *Proceedings of the IEEE International Conference on Robotics and Automation*, 1996, pp. 963–968.
312. Sharma, S.P., Kumar, D., Kumar, A., Reliability analysis of complex multi-robotic system using GA and fuzzy methodology, *Applied Soft Computing*, Vol. 12, 2012, pp. 405–415.
313. Sharp, R., Decreton, M., Radiation tolerance of components and material in nuclear robot applications, *Reliability Engineering and System Safety*, Vol. 53, 1996, pp. 291–299.
314. Sheehy, N.P., Chapman, A.J., Safety of CNC and robot technology, *Journal of Occupational Accidents*, Vol. 10, No. 1, 1988, pp. 21–28.
315. Shi, T. et al., Approach to fault on-line detection and diagnosis based on neural networks for robot in FMS, *Chinese Journal of Mechanical Engineering (English Edition)*, Vol. 11, No. 2, 1998, pp. 115–121.
316. Shima, D., Safety control on introduction of industrial robots to factories, *Safety*, Vol. 33, 1982, pp. 18–19.
317. Shulman, H.G., Olex, M.B., Designing the user-friendly robot, *Human Factors*, Vol. 27, 1985, pp. 91–98.
318. Smrcek, J., Neupauer, R., Testing of intelligent robots, development and experience, *Proceedings of the IEEE International Conference on Intelligent Engineering Systems*, 1997, pp. 119–121.
319. Sneckenberger, J.E., Etherton, J.R., Computer related hazard control needs for safe robot operations, *Proceedings of the American Society of Mechanical Engineers (ASME) International Computers in Engineering Conference*, 1983, pp. 110–113.
320. Stancliff, S.B., Dolan, J., Trebi-Ollennu, A., Planning to fail-reliability needs to be considered a priori in multirobot task allocation, *Proceedings of the IEEE International Conference on Systems, Man, and Cybernetics*, 2009, pp. 2362–2367.
321. Stancliff, S.B., Dolan, J.M., Planning to fail-using reliability to improve multirobot task allocation, *Proceedings of the SPIE Conference on Unattended Ground, Sea, and Air Sensor Technologies and Applications*, Vol. 7693, 2010, pp. 1–9.
322. Stormont, D.P., Allan, V.H., Managing risk in disaster scenarios with autonomous robots, *Journal of Systemics, Cybernetics & Informatics*, Vol. 7, No. 4, 2009, pp. 66–71.
323. Stowe, W.W., Robots, safe or hazardous?, *Professional Safety*, Vol. 28, 1983, pp. 32–35.
324. Stubbs, K., Hinds, P., Wettergreen, D., Challenges to grounding in human-robot collaboration: Errors and miscommunications in remote exploration robotics, Report No. CMU-RI-TR-06-32, Robotics Institute, Carnegie Mellon University, Pittsburgh, Pennsylvania, July 2006, pp. 1–17.
325. Sugimoto, N., Introduction to safety measures for industrial robots, *Safety*, Vol. 33, 1982, pp. 14–17.
326. Sugimoto, N., Kawaguchi, K., Fault tree analysis of hazards created by robot, *Proceedings of the 13th International Symposium on Industrial Robots and Robot 7*, 1983, pp. 9.13–9.28.

327. Sugimoto, N., Systematic robot related accidents and standardization of safety measures for robots, *Proceedings of the 14th International Symposium on Industrial Robots*, 1984, pp. 131–138.
328. Suita, K. et al., Failure-to-safety "Kyozon" system with simple contact detection and stop capabilities for safe human autonomous robot coexistence, *Proceedings of the IEEE International Conference on Robotics and Automation*, 1995, pp. 3089–3096.
329. Sun, U., Sneckenberger, J., Human-robot symbotic system safety "Criteria" for manned space flight, *Proceedings of the Annual Pittsburgh Conference: Control, Robotics, Systems, Power and Mechanical Modeling and Simulation*, Vol. 21, 1992, pp. 2327–2335.
330. Suri, R., Quantitative techniques for robotic systems analysis, in *Handbook of Industrial Robotics*, ed. S.Y. Nof, John Wiley and Sons, New York, 1985, pp. 605–638.
331. Tadokoro, S. et al., Stochastic prediction of human motion and control of robots in the service of human, *Proceedings of the International Conference on Systems, Man and Cybernetics*, 1993, pp. 503–508.
332. Takeda, T., Hirata, Y., Kosuge, K., HMM-based error recovery of dance step selection for dance partner robot, *Proceedings of the IEEE International Conference on Robotics and Automation*, 2007, pp. 1768–1773.
333. Taogeng, Z., Dingguo, S., Qun, H., Application of Bayes data fusion in robot reliability assessment, *Proceedings of the International Workshop on Bio-Robotics and Teleoperation*, 2001, pp. 262–265.
334. Tian, J., Sneckenberger, J., Performance evaluation of three pressure mats as robot work-station safety devices, in *Ergonomics of Hybrid Automated Systems*, ed. W. Karwowski, Elsevier, Amsterdam, 1988, pp. 559–565.
335. Ting, Y., Tosunoglu, S., Fernandez, B., Control algorithm for fault tolerant robots, *Proceedings of the IEEE International Conference on Robotics and Automation*, 1994, pp. 910–915.
336. Tinos, R., Navarro-Serment, L.E., Paredis, C.J.J., Fault tolerant localization for teams of distributed robots, *Proceedings of the International Conference on Intelligent Robot and Systems*, 2001, pp. 1061–1066.
337. Toye, G., Leifer, L.J., Helenic fault tolerance for robots, *Computers & Electrical Engineering*, Vol. 20, No. 6, 1994, pp. 479–497.
338. Troccaz, J., Passive arm with dynamic constraints: A solution to safety problems in medical robotics?, *Proceedings of the IEEE International Conference on Systems, Man, and Cybernetics*, 1993, pp. 166–171.
339. Tso, K., Backes, P., Fail-safe tele-autonomous robotic system for nuclear facilities, *Robotics and Computer-Integrated Manufacturing*, Vol. 10, No. 6, 1993, pp. 423–428.
340. Ukidve, C.S., McInroy, J.E., Jafari, F., Using redundancy to optimize manipulability of Stewart platforms, *IEEE/ASME Transactions on Mechatronics*, Vol. 13, No. 4, 2008, pp. 475–479.
341. Van Deest, R., Robotic safety: A potential crisis, *Professional Safety*, Vol. 29, 1984, pp. 40–42.
342. Vanderperre, E.J., Makhanov, S.S., Risk analysis of a robot-safety device system, *International Journal of Reliability, Quality and Safety Engineering*, Vol. 9, No. 1, 2002, pp. 79–87.

343. Wakita, Y. et al., Realization of safety in a coexistent robotic system by information sharing, *Proceedings of the IEEE International Conference on Robotics and Automation*, 1998, pp. 3474–3479.
344. Walker, I.D., Cavallaro, J.R., The use of fault-trees for the design of robots for hazardous environments, *Proceedings of the Annual Reliability and Maintainability Symposium*, 1996, pp. 229–235.
345. Wang, T. et al., Fuzzy reliability design of robot parts based on Weibull and normal distributions, *Proceedings of the International Conference on Computer Science and Software Engineering*, 2008, pp. 1081–1084.
346. Ward, G.R., Safety in robot control systems, *Measurement and Control*, Vol. 21, No. 9, 1988, pp. 266–271.
347. Wei, Y. et al., Visual error augmentation for enhancing motor learning and rehabilitative relearning, *Proceedings of the International Conference on Rehabilitation Robotics*, 2005, pp. 505–510.
348. Wells, D.J., Hazardous area robotics for nuclear systems maintenance: A challenge in reliability, *Proceedings of the Portland International Conference on Management of Engineering and Technology*, 1991, pp. 371–373.
349. Wells, D.J., Krishnaswami, K., Fault analysis and recovery strategies for deep-sea robots, *Proceedings of the 11th Annual Energy Sources Technology Conference and Exhibition*, 1988, pp. 39–47.
350. Wewerinke, P.H., Modeling human operator involvement in robotic systems, *Proceedings of the IEEE International Conference on Systems*, 1991, pp. 1225–1230.
351. Wikman, T.S., Branicky, M.S., Newman, W.S., Reflex control for robot system preservation, reliability and autonomy, *Computers & Electrical Engineering*, Vol. 20, No. 5, 1994, pp. 391–407.
352. Wikman, T.S., Newman, W.S., Reflex control for safe autonomous robot operation, *Reliability Engineering and System Safety*, Vol. 53, 1996, pp. 339- 347.
353. Willetts, N., *Jaguar. A* robot user's experience, *Industrial Robot*, Vol. 20, No. 1, 1993, pp. 9–12.
354. Wilson, M.S., Reliability and flexibility: A mutually exclusive problem for robotic assembly?, *IEEE Transactions on Robotics and Automation*, Vol. 12, No. 2, 1996, pp. 343–347.
355. Winfield, A.F.T., Nembrini, J., Safety in numbers: Fault-tolerance in robot swarms, *International Journal of Modelling, Identification and Control*, Vol. 1, No. 1, 2006, pp. 30–37.
356. Wygant, R.M., Donaghey, C.E., Ergonomic considerations in robot selection and safety, *Proceedings of the 31st Annual Meeting of the Human Factors Society*, 1987, pp. 181–185.
357. Xu, F., Deng, Z., The reliability apportionment for X-ray inspection real time imaging pipeline robot based on fuzzy synthesis, *Proceedings of the Fifth World Congress on Intelligent Control and Automation*, 2004, pp. 4763–4767.
358. Yamada, A., Takata, S., Reliability improvement of industrial robots by optimizing operation plans based on deterioration evaluation, *CIRP Annals–Manufacturing Technology*, Vol. 51, No. 1, 2002, pp. 319–322.
359. Yamada, Y. et al., Failure-to-safety robot system for human–robot coexistence, *Robotics and Autonomous Systems*, Vol. 18, No. 1–2, 1996, pp. 283–291.
360. Yamada, Y. et al., FTA-based issues on securing human safety in human/robot coexistence system, *Proceedings of the IEEE International Conference on Systems, Man, and Cybernetics*, 1999, pp. 1058–1063.

361. Yamada, Y., Evaluation of human pain tolerance and its application to designing safety robot mechanism for human–robot coexistence, *Journal of Robotics and Mechatronics*, Vol. 9, 1997, pp. 65–70.
362. Yang, M. et al., Environmental modeling and obstacle avoidance of mobile robots based on laser radar, *Qinghua Daxue Xuebao/Journal of Tsinghua University*, Vol. 40, No. 7, 2000, pp. 112–116.
363. Yeh, P., Barash, S., Wysocki, E., Vision system for safe robot operation, *Proceedings of the IEEE International Conference on Robotics and Automations*, 1988, pp. 1461–1465.
364. Yokomizo, Y., Occupational and safety problems caused by increases in the robot population, in *Occupational Health and Safety in Automation and Robotics*, ed. K. Noro, Taylor & Francis, London, 1987, pp. 169–174.
365. Yong, J. et al., A combined logistic and model-based approach for fault detection and identification in a climbing robot, *Proceedings of the IEEE International Conference on Robotics and Biomimetics*, 2006, pp. 1512–1516.
366. Yu, S., McCluskey, E.J., On-line testing and recovery in TMR systems for real-time applications, *Proceedings of the International Test Conference*, 2001, pp. 240–249.
367. Zhang, C., Bai, G., Extremum response surface method of reliability analysis on two-link flexible robot manipulator, *Journal of Central South University of Technology*, Vol. 19, 2012, pp. 101–107.
368. Zhang, J., Lu, T., Fuzzy fault tree analysis of a cable painting robot, *Shanghai Jiaotong Daxue Xuebao/Journal of Shanghai Jiaotong University*, Vol. 37, No. 5, 2003, pp. 62–65.
369. Zhang, X. et al., Experimental analysis of in-pipe robot accelerated life, *Proceedings of the International Conference on Mechatronic Science*, 2011, pp. 572–575.
370. Zhongxiu, S., Reliability analysis and synthesis of robot manipulators, *Proceedings of the Annual Reliability and Maintainability Symposium*, 1994, pp. 201–205.
371. Zhou, G., Position and orientation calibration and moving reliability of the robot used in pyrometallurgy process, *Zhongnan Gongye Daxue Xuebao/Journal of Central South University of Technology*, Vol. 31, No. 6, 2000, pp. 556–560.
372. Zurada, J.G., James, H., Sensory integration in a neural network-based robot safety system, *International Journal of Human Factors in Manufacturing*, Vol. 5, No. 3, 1995, pp. 325–340.

# Index

**S**

CPSIA information can be obtained
at www.ICGtesting.com
Printed in the USA
LVHW081645211222
735702LV00005B/133

9 781498 706445

# When the Sky Falls

*A Human History of Shooting Stars*

Jade Angeles Fitton

BATSFORD

First published in the United Kingdom
in 2026 by
Batsford
43 Great Ormond Street
London
WC1N 3HZ

An imprint of B. T. Batsford Holdings Limited



ISBN 9781837330652

A CIP catalogue record for this book is available from the British Library.

10 9 8 7 6 5 4 3 2 1

Reproduction by Rival Colour Ltd, UK
Printed and bound by Dream Colour, China

This book can be ordered direct from the publisher at www.batsfordbooks.com, or try your local bookshop.

Distributed throughout the UK and Europe by Abrams & Chronicle Books, 1st Floor, 22–24 Ely Place, London EC1N 6TE and 57 rue Gaston Tessier, 75166 Paris, France

www.abramsandchronicle.co.uk
info@abramsandchronicle.co.uk

# Contents

# Glossary

**Asteroid:** A stony object, sometimes with a metallic core, that orbits the Sun.

**Asteroid Belt**: A ring of debris located between Mars and Jupiter that is the home of millions of asteroids. It is the only asteroid belt in our Solar System, though other planetary systems have multiple asteroid belts.

**Bolide:** An extremely bright meteor that typically explodes within Earth's atmosphere.

**Broom star:** Archaic Chinese term for comet.

**Bushy star:** Archaic Chinese term for comet.

**Chaman tz 'utan:** An ancient Mayan term for a small comet, translating to 'a magician's cigar'.

**Chondrite**: A stony meteorite.

**Comet**: A celestial body that contains ice and dust; when near the Sun a 'tail' of gas and dust becomes visible.

**Cometa:** The Latin for comet used in medieval Europe, borrowed from the Greek *komētēs,* which means 'long-haired'.

**Elliptical orbit:** The egg-shaped trajectory of a celestial object as a consequence of gravitational forces.

**Ejecta:** Matter that is ejected, typically as a result of a meteorite impact, stellar explosion, or volcanic eruption.

**Exosphere:** The outermost (fifth) layer of Earth's atmosphere composed of light gases.

**Flaming sword:** Medieval European term for a comet.

**Flower of the sky:** A Victorian poetic term for a comet.

**Guest star:** Ancient Chinese term for a comet or supernova.

**Hyperbolic orbit**: A one-time flight path of a celestial object that does not repeat or return.

**Impact event:** When an object from space, such as an asteroid or meteor, collides with a planet or its satellites.

**Impact winter:** The theorised period of prolonged global cooling after an asteroid or comet impact. The resulting dust, debris and soot block sunlight for many months or years, resulting in reduced photosynthesis and ultimate starvation for many species.

**Kamacite:** A nickel-iron alloy (up to 7% nickel) primarily found in iron meteorites and occasionally stony meteorites.

**Kuiper Belt:** A disk-shaped home of relics from the formation of the Solar System, containing comets and Pluto, located beyond Neptune. It differs from the Asteroid Belt in size (much larger), location, and contains bodies largely comprised of ice and gas.

**Long-period comet:** Orbital periods can be many hundreds, thousands or even millions of years and originate from the Oort Cloud.

**Mesosphere:** The third and coldest layer of Earth's atmosphere, where friction with the layer's gas molecules causes meteors burn up.

**Meteor:** When a meteoroid enters Earth's atmosphere, it heats up and vaporises due to friction with the air, creating a bright streak of light in the sky known as a meteor or 'shooting star'.

**Meteorite:** If a meteoroid is large enough, it can survive its fiery passage through the atmosphere and strike the Earth's surface. The recovered object is then called a meteorite.

**Meteoroid:** The general term for a small, rocky or metallic object traveling through space.

**Meteor air burst:** The explosion of a meteor upon entry to a planet's atmosphere.

**Meteor shower:** The debris of comets or asteroids; there are nine major meteor showers and 21 minor meteor showers every year; they are named after constellations from which they are deemed to fall.

**Octahedrite:** The most common iron meteorite.

**Oort Cloud:** A theoretical ring around the Solar System that is the home of long-period comets.

**Pallasites**: Stony meteorites with iron-nickel matrix that contains yellow-green olivine silicate crystals.

**Parallax:** The apparent shift of an object when viewed from different points of the Earth's orbit.

**Perihelion:** The closest a planet, asteroid or comet gets to the Sun, typically making this the point the object is at its brightest.

**Pheasant star:** Archaic Chinese term for a shooting star.

**Primitive comet/asteroid:** A body as ancient as our Solar System, containing residual materials of its creation.

**Pyrite:** In this instance: meteorites containing iron sulphide or 'fool's gold'.

**Shooting star:** A rapidly moving celestial object, typically a meteor but often associated with other fast-moving materials in space.

**Short-period comet:** Orbital periods of under 200 years and originate from the Kuiper Belt.

**Stratosphere**: The second layer of the atmosphere, which contains the ozone layer.

**Strewn Field:** Area of dispersal of a meteorite fall.

**Sungrazer:** A comet that passes very close to the Sun.

**Superbolide:** An exceptionally bright meteor, or fireball, visible in daylight, they are around 100 times brighter than a full Moon.

**Taenite:** A mineral alloy of iron and nickel (nickel proportions of up to 65%) found mostly in iron meteorites.

**Tektite**: A black, molten glassy ejecta from a meteorite impact.

**Thermosphere:** The fourth layer of Earth's atmosphere and the layer where aurora borealis and aurora australis occur.

**Thunderstone:** Colloquial term for the Ensisheim meteorite of 1492.

**Troposphere:** The first and lowest layer of Earth's atmosphere, which begins on the Earth's surface up to an altitude of around 7½ miles (12km).

**Uncommon Star**: Tudor term for the comet of 1577.

**Widmanstätten pattern:** The interleaving of kamacite and taenite that results in a pattern unique to each octahedrite or pallasite meteorite.

# Introduction

## L'Aigle, France – 1803

Swallows fly through an early spring afternoon in the rural French town of L'Aigle, Normandy. Horse and carts clatter down the main streets, transporting the town's mesdames, gossips and local celebrities to lunch. A clerk scratches at his paper with a quill, signing a court release. Thirty leagues away, a young boilermaker, laden with tools and sheets of metal, packs up after his last job – he whistles to his friend across the street before hopping into a light cart that speeds him back home to L'Aigle. In the fields, oxen heave ploughs through the earth, their tough white hides whipped by peasant farmers longing for their lunch.

It is April in the eleventh year of the Republican Calendar, and the French Revolution simmers in the country's memory. In Paris, it is the Age of Reason, and rational thinking has superseded luxury, superstition and frivolity: men of science dissect and discount the cosmos in the wooden rooms of the Institute of France. Unbeknown to them, in the not too distant future, they will be in receipt of hundreds of rocks reported to have fallen from the sky and onto the small town of L'Aigle, this very day. Despite their misgivings, they will dispatch a young physicist, Jean-Baptiste Biot, to the town to investigate a phenomenon still regarded by such enlightened men as the superstition of peasants; like the sky raining blood, milk or honey, the idea of rocks falling from the heavens is for the birds.

Meanwhile, back in L'Aigle, it is now approaching 1 o'clock. The farmer climbs onto his plough to take an early lunch. Sweating, he swills some beer and, tilting his head back to reach the dregs, catches a series of bright flashes, brighter even than the clear spring sky. He blinks. More flashes, so dazzling that when he rubs his eyes streaks of colour glow across his retinas. He hears a sound like a whistle, and a small rock lands by his feet. He turns to see who has thrown it. The field is empty but he can hear shouts in the distance. Then comes a rushing roar, like a mortar.

The townspeople of L'Aigle have grown up with the tale of a rock that had fallen from the sky some three centuries earlier, onto what was then the Germanic town of Ensisheim in Alsace, which later became part of north-east France. In 1492, a 280lb (127kg) meteorite fell to Earth. The 'thunderstone' left a hole in the ground the size of a small child and foretold the defeat of French forces at the hands of the Holy Roman Emperor that same year. It had been kept chained inside the town's church so that it would not 'wander' until the recent Revolution, when it was liberated along with much of the church's property by revolutionaries. To symbolise the triumph of the secular Revolution over religion and superstition, the meteorite had been temporarily relocated to Colmar, on France's border with Germany, to undergo scientific analysis.

Like most people, the farmer had heard the legend of the thunderstone. When he was a boy, his grandfather had told him that the victorious Holy Roman Emperor had kept fragments of it for good luck. It had been a source of pride for the Germanic peoples, a talismanic weapon against the French. In their oral history, stones falling from the heavens foreshadowed plague, pestilence and defeat. In the field the falling rock is accompanied by the smell of sulphur – the smell of hell. The crack of an almighty explosion is followed by a sound like a large tree falling, and the sky now belches a torrent of rocks. The farmer hides beneath his plough, inhaling the smell of damp earth as the trailing flashes that he had once marvelled at as a boy transform into a globe of fire before his eyes.

Two hundred and twenty-three years later, a dark grey rock, smoothed as if by a river of time and space, sits under white LED lights inside a glass box in a provincial museum in the sleepy town of Angers, in western France. If a rock were ever to be described as looking mischievous, it would be the L'Aigle meteorite.

The L'Aigle meteorite was one of over 3,000 rocks that fell to Earth that April day in 1803. Originating from a larger meteor that was dislodged from the asteroid belt – a necklace of frigid debris between Mars and Jupiter – this meteorite travelled over 92 million miles (148 million km) to change our understanding of shooting stars forever.

Because the strewn field (the area over which meteorites from a single fall are dispersed) was so vast, eye-witness accounts of the event weren't limited to the tales of isolated peasant farmers, as had so often been the case previously, and whose accounts were typically discounted and ridiculed. Now, educated, upper-class citizens, who had nothing to gain from perceived fantasising, were giving the same testimonial as peasants to Jean-Baptise Biot, the young physicist that had been dispatched by the Institute.

Accumulating testimony from all walks of life – farmers, dignitaries, clerks and young boilermakers heading home, all talking about the 'rain of stones thrown by the meteor' – Jean Baptiste Biot would be able to prove to rational men that rocks falling from the sky were no longer a matter of superstition and folk belief, but a reality.

Culturally, instinctively, how much did this shift in perception and understanding matter to the Western mind? Did scientific understanding of these celestial phenomena shift our association with shooting stars away from the portentous or the miraculous? Were there other cultures that already saw them as gifts from the gods, and used these cosmic offerings to better their day-to-day lives? What perceptions, superstitions and beliefs have we held historically regarding shooting stars across the globe – have they always been portents of doom, signs of the falling of kings, the physical manifestation of a miracle? Or were there other, stranger beliefs regarding these uncommon stars?

And did this new, rational knowledge change how a person – standing on a moor, in a desert, out at sea, driving home – felt when they saw a light tearing through the sky? Did we continue to feel as we have always done when witnessing a shooting star, filled with awe and terror in equal measure? *When the Sky Falls* will delve into our ancient past, turn over our present and grasp at our potential futures to explore where shooting stars have taken us, and where they may yet lead.

*

In the not so distant past, I grew up near a Dark Sky Zone, which ensured our evenings in the countryside were ruled by the stars, or the clouds that obscured them, depending on which way the wind was blowing. I remember nights during the long summer holidays where I lay out on our hard, cold, stony drive, staring like a thwarted magpie at the glittering sky, waiting and watching for a lone shooting star or the volley fire of the Perseids. This was before there was the clutter of satellites that now haunt our sky, which hardly detract from the surprise of a shooting star but can become something of a frustration, causing the observing hand to shoot up and point, and then retract, disappointed. It was also before I had any real concept of, or interest in, what these declining lights were aside from the primal enchantment they held for a young girl whose principal interest was in the mud and all that sprang from it. In place of the fireflies and glowworms I read about and longed to see, I had these bright beings. I looked up simply because I could, because this impossible canopy hung above me.

My fascination with shooting stars endured a punk rock phase and a move to a city; but it was then largely confined to the blue glow of a computer screen as I learned about the backrooms of heaven, where such superlunary phenomena originate. Where there is no life as we know it. Where all is an inanimate kingdom of rocks waiting to be nudged, like chicks nesting on a cliff, into their first and only flight, their moment in the Sun.

Confused about where I was and what I was doing with my life, I watched plasma, gases and debris fight and play; rise and fall like empires; explode in furious death rages; swallow themselves whole. Travelling to Los Angeles during this period, I found that there were places, even within a city, where residents were encouraged to turn off their outdoor lights at night to prevent light pollution and increase the visibility of stars. And while I looked up and watched the sky's white lights, infinite as the white lights on the eight-lane highways, I wondered why this phenomenon wasn't more prevalent in towns and cities in the UK. I was envious of the residents of Topanga Canyon, and the longer I lived in a city the more I felt that some priceless inheritance was missing in a life without a clear night sky.

By the time I returned to the countryside and its open-ended evenings, I had changed careers and started to write. I developed a fascination, not only with meteors and comets, but with asteroids, their trajectories and their gravitational pull, together with a morbid excitement when entertaining their absolution. I spent four years writing a novel that, in part, followed the journey of an impending asteroid. The book was as unwieldy as the asteroid that featured in it and would never be finished, but the research I did then informed and perpetuated this enchantment with fires that fly.

I have since spent time speaking with astrophysicists, read papers and sifted through archeological and anthropological research about humankind's relationship with, and beliefs surrounding, sky fires; and I have sought out more dark zones to better watch the skies fall. But the most memorable occasion was the time I was not looking for shooting stars, when I found myself on a tiny island in lockdown during the Geminid meteor shower – one of only two meteor showers created by the debris of an asteroid rather than a comet. I wrapped myself up and sat with a glass of wine in a deckchair in December, outside the island church I was living in, and watched as sparks seemingly fell from the universe, only to be extinguished by the howling wind.

Certainly, this was my most memorable occasion – until last night, the day before I was due to complete this book. Driving home on potholed country back roads, I looked up above the dark fields and evergreen conifers at what, at first, appeared to be a lone firework above the forest. Then I realised what I was seeing: a meteor falling from north-east to south-west. A long, smoky orange tail chased the head as it barrelled down like an impending, fire-starting fist. I stopped the car and watched open-mouthed as its brightness increased, becoming a flare of white-green, before disappearing, leaving the crisp night black and silent, as if it had never been there. It was a rogue, an imp in the uniform march of time. A sight I had longed for my whole life. In that instant there was no ambiguity between me and the universe. Time eroded and I experienced the full grace of existence: the present moment.

*When the Sky Falls* is the culmination of these years and experiences: I hope to share the curious, the terrifying, the beautiful, the strange, the dark, the backwards, the progressive and the inspiration we have drawn from such rare, travelling stars. Meteors, meteorites, meteoroids, bolides, comets and asteroids have shaped our belief systems, folklores, countries, monarchies, histories, our understanding of the universe, and the course of life itself. Perhaps what is most remarkable is that, after billions of years, they continue to influence many of these things.

What I find most heartening about our shared history of shooting stars is not their propensity to bring about sublime realisations from the exceptionally gifted, it is that we have not needed high intellect to see the miracle or the doom. It was the labourers out on the land who saw stones fall from the sky and needed no further evidence to comprehend that, strange as it may be, this was the way of things. It was the person, anywhere in the world, who picked up a rock that had fallen from the sky and knew, instinctively, that it was different to the stone they stood upon. And for a time, they alone knew that we could reach out and touch something that was out of this world.

Devon, England, 26 November 2025

# PART ONE

# When the City is a Star:

## *Art, Culture and the Shooting Star*

# Melancholia

On the wall of my study there is a portrait of a woman who sits with her chin in her hand, slumped after a day's effort to capture what lies perpetually beyond her. By her side is a devoted hound, curled up, one eye half-open, sensing that something approaches. At her feet is an orb, a polyhedron with a spectre on one of its faces, a large, serrated blade and a wooden plane. Above her is a mathematical table, known as a 'magic square', and in the centre of its bottom row are the numbers 1514: the year the German artist Albrecht Dürer created the work. Out of the woman's window, a bright meteor or comet is encroaching beneath the arch of a rainbow, or a moonbow – depending on the perceived hour. At her back hang a pair of wings. A dejected putto, a cherub-like infant, sits on a wooden wheel by her side – mirroring the female figure: perhaps she has always been like this. The woman's gaze lies not on her page but outside, on the watery horizon, on the bat clasping a streamer as it smashes into the scene, heralding the blazing star that remains beyond the woman's awareness for now. The female figure is the personification of melancholy: she is the creative being that taps at the boundaries of the high-mind, searching for a way in.

Having toiled over the same idea for months, or years, it is understandable that one might reside in the doldrums of low spirits and dejection from time to time, but *Melencolia I* was painted at a time when it was believed that melancholy was in the blood. In medieval thought, melancholy was one of the four temperaments that corresponded with the four humours of the body, and those of a melancholy disposition were thought to be of a quiet and analytic temperament. The others were: sanguine (optimistic, sociable); choleric (irritable, short-tempered); and phlegmatic (calm, relaxed). Melancholy was believed to be both the malaise that affects the intellect, and the source of creative inspiration. Associated with black bile, it corresponded with cold and dryness, and the planet Saturn, which some have suggested the meteor may represent.

Melancholy manifested in persisting sadness or depression, fear, anxiety, irritability and a sense of hopelessness, as well as physical symptoms such as poor appetite, sleeplessness and delusions. As the sufferer rolled between swells of rapture and resignation, these symptoms were seen as the other side of the coin, the debt that comes with the gift of creativity – for the wealthy, powerful or the influential. If associated with a man, melancholy could be perceived as confirmation of genius, 'the divine frenzy'. If it were manifest in a woman, she was likely to be diagnosed with 'hysteria' instead, as the symptoms often overlapped.

For Dürer to depict the *homo literatus* – the savant – as a female figure on her island of genius was ground-breaking – not only in regards to her sex, but in depicting melancholy as a battle of intellect rather than a state of intellectual paralysis. This work would influence how artists and writers depicted melancholy in their future works, the most famous being the German artist Lucas Cranach the Elder's four depictions of melancholy (between 1528 –1533) which personify the condition as a female figure with wings in a similarly slumped pose.

Decoding the many hidden messages in *Melencolia I* has occupied art historians for centuries. As it is widely believed that the engraving symbolizes that a melancholy disposition is the reluctant bridesmaid of creativity, which had been specifically associated with writers, artists and thinkers since Aristotle, this has led some to suppose that the image is, in fact, a spiritual self-portrait of the artist himself (as almost all artworks inevitably are, to one degree or another).

During his lifetime, Dürer will have witnessed a comet in 1491, heard about the fireball that was the Ensisheim meteorite in 1492, and witnessed another large meteorite fall in 1493. The Ensisheim meteorite, or 'thunderstone', which landed in a wheat field and was chained to the walled town's church until the French Revolution, was the most famous of these instances. Its impact was heard over 100 miles (160km) away and it was depicted in a woodcut as a boulder chasing bands of fire through a cowl of cloud above the town in the *Nuremberg Chronicle* (1493) – it is regarded as the most likely inspiration for the shooting star in Dürer's engraving. The Ensisheim meteorite is thought to have influenced Dürer profoundly, and a depiction of the Ensisheim meteorite is also believed to appear on the reverse panel of *Saint Jerome in the Wilderness* (1496).

In turn, Lars von Trier's film *Melancholia* (2011) was profoundly influenced by Dürer's *Melencolia I*, with the film's title and themes directly inspired by Dürer's depiction of melancholy, and the artwork's suggestion of an impending celestial body. Reflecting the events of von Trier's film, at the time Dürer lived, comets and meteorites were generally accepted as harbingers of a cosmic dread – an End Time – that was persistent in medieval religious thought. But, when I look at this image, I like to think that the meteor represents a desired climax: the woman has been left alone long enough to transcend the boundaries of thought – she is on the precipice of breaking through. It is a breakthrough that, I imagine, is symbolised by the meteor, just out of sight, just entering the outermost periphery of her vision.

This flash of inspiration emerges from the seemingly endless black of our unknowing, tearing over the worn and tangled fields of repetitive thought, burning through all other matters, and illuminating new territory in its wake – swaling an unchartered path of contemplation, so that we might stumble upon some little revelation. In both our distant and more recent past, shooting stars have been the precursors of intellectual and creative breakthroughs. From the lore and superstition of the Dark Ages, to the astronomical debate surrounding these phenomena in the Early Modern Period onwards (14th–19th centuries), artists, writers, scientists, alchemists, directors, astronomers, composers and even city planners have drawn upon the legends and discoveries inspired by shooting stars. Over the centuries, as the mist surrounding the composition and origins of shooting stars gradually faded, like the bloom on a grape wiped clean, their role as a harbinger of inspiration appeared only to crystallise on the surface of the creative psyche. They continue to represent reaches where the mind cannot throw itself, where calculations cannot be balanced through conscious thought.

# The School of Night

Walking out one night into the pooling black, a tired actor rubs his face and removes his cap, smoothing its long heron feather in ritual conclusion of the evening's work. He takes a breath of the deep autumn air and steps out onto the banks of the wide river, listening as a sloshing shanty of oars ferry passengers to Mortlake. Gazing at the twinkling pin-pricks in the inky canopy, a star with a burning taper catches his eye. It is night in 1592 and now it is time for the actor to write.

Shakespeare lived during the beginning of the end of the Earth as we once knew it, as the centre of the known Universe, and lived at the beginning of the heliocentric theory first proposed by the Polish astronomer and polymath Nicolaus Copernicus in 1543, which placed the Sun at the centre. Because of this, Copernicus was widely despised in Western Europe. In the Christian world, heliocentrism was generally deemed dangerous, radical and even heretical, as it contradicted religious ideas that the Earth was stationary. 'He has fixed the earth firm,' the Bible says in Psalms, 'immovable'.

There were two kingdoms that not only permitted such freedom of thought, but actively encouraged it, thus the courts of Elizabeth I of England and the Habsburg empire of Rudolf II were duly rewarded with the presence of some of the brightest minds of the period. The latter is where astronomers Tycho Brahe and Johannes Kepler travelled (from Denmark and Germany respectively) to study, build on and debate Copernicus's earlier work. The Elizabethan court, on the other hand, welcomed a man called Dr Dee.

John Dee has become renowned for his influence on astrology and astronomy. (He also conducted less successful forays into divining tools, such as obsidian scrying stones, and claimed that, through a collaboration with a medium-cum-charlatan called Edward Kelley, they had received angelic commands and transcribed them in 'Enochian', the supposed language of the angels. It was in this language that Kelley happened to receive the angelic command he claimed entitled him to a night with Dee's wife.) Dee studied the Copernican universal models and ensured they were spread rhizomatically, establishing the astronomical debate as a source of glittering curiosity to the intellectuals of an increasingly inquisitive Elizabethan society. As Dee was one of the Queen's closest confidants and advisors, the parameters of thought became all the more permissive under his influence, extending into the far reaches of the firmament.

These intellectual boundaries were tested by the so-called School of Night, a circle of radical thinkers, writers and courtiers. In 1592, a Jesuit priest had referred to the groups as the School of Atheism; the name 'School of Night' was coined later in 1903 by author Arthur Acheson, derived from a line in William Shakespeare's *Love's Labours Lost* (c.1596), which was perhaps referencing the priest's denouncement of the group.

*O paradox! Black is the badge of hell,*
*The hue of dungeons and the school of night …*

On the heels of the enlightened astronomical debate, some of the monarch's subjects were starting to look at the heavens with a new, and more penetrating gaze. Henry VIII's earlier reforms had made the monarch Supreme Head of the Church of England, which meant that this was a time when atheism could carry the same penalty as treason (death). All the supposed members of the group were suspected of atheism so, despite the intellectually permissive atmosphere, caution was still advised. This is perhaps why there is no written evidence that the School of Night existed in any formal capacity, but it is believed to have comprised a number of learned members, including the Earl of Northumberland, known as 'the Wizard Earl'; explorer Sir Walter Raleigh; the poet Sir Philip Sidney; the scientist and astronomer Thomas Harriott – who was in secret correspondence with Johannes Kepler; the Earl of Southampton; the linguist John Florio; and the poet and playwright Christopher Marlowe.

It is now generally accepted that Shakespeare collaborated with Marlowe on the plays that comprise the trilogy of *Henry VI*. The Earl of Southampton, Henry Wriothesley (son of Thomas Wriothesley or 'Call-Me' for fans of *Wolf Hall*), was Shakespeare's patron and John Florio was his secretary. Therefore, it seems almost impossible that radical ideas about the heavens did not find their way to the man whose genius was a thief, and whose works are a collection of all strange, beautiful and damnable things.

Perhaps the most culturally significant idea to arise from this scene is that, as far as we know, it is Shakespeare who first conjured the expression 'shooting star' as being synonymous with a meteor. He writes in *Richard II*:

*I see thy glory like a shooting star*
*Fall to the base earth from the firmament.*

# The Missing Piece

On a sheet of beige woven paper a shooting star oscillates, subsuming most of the page with its ignited tails; there is just room still for a smaller, spider-like, timid shooting star, spiralling in the corner.

The French-American artist and sculptor Louise Bourgeois' most notable depiction of shooting stars is in the dry point work *Shooting Stars* (1989). They also feature in a series of prints made in the portfolio *Anatomy*, which features a print called *Etoiles*, a study of stars, including a shooting star. A self-professed seeker and searcher her whole life, a shooting star might be seen as a momentary completion of the search; transitory, apparently from nothing, returning to nothing.

In *Shooting Stars* her star takes a spiral form, one of her preferred abstract shapes, denoting both control and liberation, for we cannot have one without the other. Spirals are deemed to symbolise spiritual evolution, a connection to the divine, rebirth and the oscillating manner in which we advance and recede from our own lives. Of this work, Bourgeois said, 'The spiral has reached too far … it is limitless … pie in the sky. It bursts … it explodes … you lose everything.'

But what if you lost something that felt like everything, and you never quite knew what it was? This was how Bourgeois felt, and why she was propelled to spend her life searching for her 'missing piece'.

Bourgeois' search for that which was missing from the whole, her desire to capture the invisible, to transmute the intangible into something a hand could hold, likely stems from childhood. At a young age, Bourgeois worked with her parents in the tapestry restoration business. It was her job to draw the missing pieces of fabric that needed to be rewoven and restored – something her most common motif, a spider, is renowned for. 'If you bash into the web of a spider,' Bourgeois said, 'she doesn't get mad. She weaves and repairs it.'

Many of us may sympathise with Bourgeois when it comes to the sense of something 'missing'; we rarely know what it is, we will likely never know, all we have is the sensation of a persisting absence of … something. This can drive us to doom, or it can push us to seek it out, through creation, through the snatching of things from the sky. Driven to disaster or to the stars, in moments of awe, we are unified, as that missing piece is returned to us.

# White Star

It was gin-clear on 14 April 1912, the evening the *Titanic* began to sink, and the night of a solar storm. Aurora borealis rippled across the horizon like fluid in a lava lamp. Some have argued that the luminous streaks reported by survivors may have been meteors as well as aurora, as the date falls just before the annual arrival of the Lyrid meteor shower. Perhaps they witnessed both.

These reports of glowing lights and trailing stars led director James Cameron to film a scene featuring shooting stars in his film, *Titanic* (1997). It was eventually cut from the final edit, but deleted scenes show Rose (Kate Winslet) making pronouncements about how 'vast and endless' the universe is, 'and how small we are', as she is reluctant to leave poor Jack (Leonardo DiCaprio) on the deck to return to her first-class cabin. As the two lovers discuss Rose's uniqueness among her superficial, socialite crowd, a long shooting star streaks over the ship through the crisp, Arctic sky, and Jack, full of Wisconsin country charm, says, 'You know, my pops used to tell me, that every time you saw one, it was a soul going to heaven.' Rose decides it is prudent not to wish upon the shooting star as she would only wish for 'something she can't have' and, getting chilly, she hurries back to her cosy five-star compartment.

Later, Cameron explained that the motif of stars in the film was cut for brevity's sake, but the scene in which Jack tells Rose a shooting star is 'a soul going to heaven' was intended to be part of that larger plan symbolising what Western cultures have long since come to associate with shootings stars: fate and hope.

Viewed from a frozen lifeboat in the reality of that numb-cold night of 1912, between the fixed and fierce blue stars, it may have been the ones that fell that inspired the survivor, Dr Washington Dodge, to write this line in his account of the fate of the White Star liner: 'We observed the closing incidents the gradual submergence of the ship forward – The final extinguishment suddenly of all her lights – The final plunge downward as a shooting star full from the Zenith visible nearly to the horizon.'

# The Messenger

It's a scene we've seen a thousand times – more, perhaps. A group of disparate characters marvelling at a baby in a humble manger. But when you look more closely at Bartolomé Murillo's *Adoration of the Magi* (c.1660), cast your gaze away from all the adoring and into the sky above, things become a little more interesting.

Influential celestial reports from the 17th century, such as Galileo's *The Starry Messenger* and *The Assayer*, which supported the Copernican heliocentric model, but also incorrectly argued that shooting stars were heavenly phantoms rather than physical phenomena, were influential to writers and artists of the day. Following these publications, Galileo went on to conduct a best-selling debate with Orazio Grassi (writing under the pseudonym of Lothario Sarsi), a 'nobody' Jesuit priest who argued correctly that comets were, in fact, physical entities.

This heated debate, which lasted several years, encouraged artists to reconsider the cosmos, and speculate that the star followed by the Magi was a shooting star, with Spanish Baroque painter Murillo showing it as a comet. Revolutionary as it may seem, Murillo was not the trailblazer in this regard. In 1304, long before Murillo and shortly after a return of Halley's Comet, the Italian artist Giotto depicted the star of Bethlehem as a comet. He thereby contradicted the general consensus of the time that comets were messengers of God's displeasure. In Giotto's painting, the comet is still a divine emissary, but on this occasion it comes with a different message. And not without a certain degree of supporting evidence.

Chinese records from the Han dynasty show that there was a 'broom star', or comet, in the year 5 BCE, which corresponds to what biblical scholars believe to be the actual year of Jesus's birth. So, the star of Bethlehem may well have been a comet. Those who believe this theory support it with evidence from the Bible, with the star said to have been seen 'stopping' above Bethlehem, translated in the King James Bible (1611) as having 'stood' over Bethlehem, which was common parlance at the time for the apparently slow flight of a comet. Likewise, Halley's

Comet was described by Roman historian, Cassius Dio, as having 'stood' over Rome in 12 BCE before the death of Roman general, Marcus Agrippa.

The Magi are known as the 'wise men', and it has been supposed that they may have been astrologers, as they were prompted to follow some strange and distant star. Divining meaning from the stars was adhering to Pagan custom but, conveniently, this particular star was not viewed as a natural phenomenon and rather a light created by God himself. In Murillo's painting, above the magi and the crude shelter, viewed through a clearing in the bank of cloud, the luminous star burns, its white-hot tail clearly visible. In the artist's mind, it symbolises the divine timing of a marvel of nature, and it is in the ascent.

# Christmas Eve

It was the night before Christmas and a witch had stolen a sleeveful of stars from the crisp Ukrainian sky. Shortly afterwards, the Devil scorched his left hand on the Moon, next, his right. Then the Devil steeled himself and, with both hands, he seized our bony satellite from the sky.

This is how we enter the story 'The Night of Christmas Eve' by the Russian writer, Nikolai Gogol. Written in 1860, the tale is inspired by pantheism and Russian folk beliefs, as well as a good old-fashioned Christian Devil, because it would be unchristian not to include him in the Christmas festivities. It is the story of a blacksmith in love and his fight to overpower the Devil. In one scene, the

blacksmith, Vakula, is taken on a ride through the night sky by the Devil to the wintery land of St Petersburg.

Gogol, and this short story in particular, inspired Russian writers and composers, including Dostoyevsky, Tchaikovsky and Nikolai Rimsky-Korsakov. The latter was sufficiently inspired by Vakula's journey through the sky to compose two interludes in his opera *Christmas Eve*, and to name those interludes in honour of sky fires.

You may not be familiar with the name Rimsky-Korsakov, but you are likely to have heard his famous interlude 'Flight of the Bumblebee'. He was a turn-of-the-century composer, Igor Stravinsky's teacher, and member of The Five, or the Mighty Handful, a group of musicians known for their love of folk music. Like Gogol, Rimsky-Korsakov became renowned for his connection to Russian folklore and for incorporating it into his compositions.

There is no mention of shooting stars in Gogol's short story, but as with 'Flight of the Bumblebee', and in accordance with folk tradition, in *Christmas Eve*, Rimsky-Korsakov's composition was influenced by the natural world around him. Whether, like Murillo, Rimsky-Korsakov was also inspired by the idea of the star of Bethlehem being a comet is unknown and, due to his pantheism, unlikely. But he will have observed some of the nine major meteor showers that are visible every year, and probably several comets during his lifetime, including one particularly spectacular sungrazing Great Comet 12 years before he composed *Christmas Eve*.

With its sweeping violins and violas, peaking piccolos, tinkling cymbals and swooping flutes, the piece 'Parade of the Comets' evokes the sounds and images of flickering stars and an arching comet as it appears coursing across the horizon through which Vakula, the blacksmith, flies. A later interlude, 'Csárdás and the Rain of Shooting Stars', evokes the more frantic falling of a meteor shower, presumably representing the Geminid meteor shower, where stars fall each December.

If the sky is cloudy, there is too much light pollution where you live to witness the Geminids, or if you are simply looking for something that will evoke the magic of Christmas Eve, you might listen to *Christmas Eve* by Nikolai Rimsky-Korsakov, and let imagination do the rest.

# Pines Where Stars Have Fallen

Legend has it that in western Japan over a thousand years ago, on 10 May in the year 861, a young boy witnessed a meteorite plummet to Earth. After he watched the star fall from above and felt it shake the ground on which he stood, he ran back to the village. This meteorite was a blessing, and he would not keep material evidence of the gods' benevolence to himself. The boy led the villagers to the impact site of the meteor, and Shinto priests from the local shrine quickly accepted the boy's story that the foreign stone had fallen from the sky. They duly preserved the meteorite as 'shrine treasure' in a wooden box. In the Shinto religion, meteors are deemed a blessing from on high, and so a new shrine was built where the meteor had landed.

With no known founder of the Shinto religion or formal scriptures, oral folklore kept rituals, legends, events and beliefs alive until they were recorded in the *Kojiki* (*Records of Ancient Matters*), composed during the early 8th century. The Shinto religion has a tradition of animism, where natural objects are believed to possess spirits. The spirit possessed by a meteorite originating from an asteroid belt at the borderlands of our Solar System is that of a traveller, as the Shinto believe our soul, our *mitama*, is destined to be.

The chondrite meteorite witnessed by the boy, now known as the Nōgata meteorite, is believed to be the oldest observed meteorite fall. And since then, a number of Shinto shrines and Buddhist temples are believed to have been established on the sites of 'fallen stars', on ground consecrated by the heavens themselves.

Today, this meteorite is preserved at the Suga Jinja shrine in Nōgata, where it fell over a thousand years ago. About the size of a pebble, the Nōgata meteorite is brought out and paraded before the public at the Grand Shrine Festival that takes place every five years. It is displayed on an ornamental cart painted red, gold and black (the colours that symbolise wealth, prestige and divine connection (red); masculinity and solemnity (gold); and protection and strength (black) in Shinto tradition). The cart is pulled by priests wearing emerald-green tunics and black admiral-like caps, and through this meteorite, the people of the Shinto religion believe they can feel the spirit of nature, the super-natural.

Through the Shinto lens, many of nature's spirits could be seen to reside in the collections of the Natural History Museum, London, including the wandering spirit of a 9lb (4kg) fragment of a bronzite-chondrite meteorite. This fragment was sold to the museum in 1883 and is one of four meteorites that fell on Ogi, on the island of Kyushu in south-east Japan, at around 11 am on the morning of 8 June 1741. These Ogi meteorites were believed to have been the stones used by Shokujo, the celestial weaver maiden of Japanese folklore, to steady her loom as she weaved, until they slipped and tumbled to Earth. The meteorites weighed a total of 31lb (14kg) and were used for many years as divine offerings to Shokujo.

The notable Nabeshima family had owned the largest meteorite from the Ogi fall, weighing over 12lb (5.6kg), but it was reportedly destroyed by American bombs in World War II. Today, Japan may only retain small physical fragments of the Ogi meteorites, but due to the sacred nature of the objects, many fragments of other meteorites are still preserved in temples, shrines and museums (including the 290lb/130kg Kesen meteorite that fell in 1850), and by the particularly robust Japanese meteor collecting community.

Sometimes a physical fragment is not required for veneration and the residual spirit of a meteorite more than suffices, existing even in place names inspired by legendary star falls, such as such as Kudamatsu, 'pine where stars fallen', in Yamaguchi Prefecture; Bisei-cho, 'beautiful star', in Okayama Prefecture; and Hoshida, 'village of star', in Osaka Prefecture.

In celebration of their connection with the stars, Bisei-cho – 'beautiful star' – has created a light pollution prevention ordinance to enable visitors to better view the night sky from the prefecture's observatory. One of the driving forces for this ordinance is the Tanabata festival, held in honour of Shokujo on the seventh day of the seventh month, when the cosmic weaver and her cowherd spouse reunite across the River of Heaven that is the Milky Way, from which the meteorites fall.

# The Brightest Point

'I came in with Halley's Comet,' Mark Twain said of his birth, in 1835. That year, the periodical comet was observed by astronomers worldwide and was visible for about a month, first appearing in August as a faint object, and developing a long tail by October. It reached its perihelion, its brightest point, in November, the month Twain was born.

In 1682, that same comet tore through the skies of Islington, London. A young astronomer watched it from his home observatory, noted the comet, and began to question whether he was experiencing a déjà vu, of sorts.

Edmund Halley's record in Islington was by no means the first time this comet had been documented. It was first recorded by Chinese astronomers in 239 BCE, calling it a 'broom star'. The comet was perhaps most famously observed in 1066, as it heralded the Norman invasion of England by William the Conqueror and the defeat of King Harold. This heralding is now immortalised in yellow and orange thread, stitched into the Bayeux Tapestry. Below the embroidered comet, courtly men cower between their towers and point at the flaming object, the text above reads 'ISTI MIRANT STELLA' ('These men wonder at the star'). But to these wondering men, the women who embroidered them, and to the many that came before them, this comet was not a single returning object, it was one of many random fires.

In Cambridge in 1682, Isaac Newton was also observing this comet. His hand-written notes have been digitised by the University of Cambridge and are available for anyone to view online. The notes are written on a torn and frayed, discoloured and flyspecked, piece of paper with fading, browning ink. The top of the page reads 'Is. Newton Anno 1682', and he has illustrated several diagrams of the comet, approximating its proximity to other stars and planets. Below these diagrams are more detailed notes, day by day, of the comet's trajectory and the times Newton observed it.

On Tuesday 22 August, Newton noted that 'The comet passed about 8 or 10 [degrees] above the star V. The tayle

[tail] was crooked, and converse side southward was sensibly brighter than the concave. The head scarce so luminous as a star of a first magnitude but more luminous than one of the second.'* The comet would reach its perihelion on 15 September 1682.

Following the comet's appearance in 1682 and Halley's continued study of it, Halley visited Newton in Cambridge in 1684, and asked if Newton knew what forces were governing this comet. Newton immediately replied that it would be an ellipsis and went to get his now-priceless notes ... but couldn't find them. He promised to send them to Halley when he did.

Newton's observations of this comet would influence his yet-to-be published *Principia Mathematica*, containing his theory of gravitation and the laws of motion, which he had formulated during the 'Plague Years' of 1665–67, but not published until Halley helped convince him to.

From his home in Islington, Halley continued to analyse the comet's path. Employing Isaac Newton's recently published theory of gravitation, Halley calculated that the comets observed in 1531, 1607 and in 1682 were not disparate entities flying through the black cosmos with abandon, but the same object travelling

* I could not find a transcription of Newton's notes, so this is my best attempt at deciphering his handwriting.

through the dark gardens of space on an elliptical orbit. Halley predicted that the comet would return in 1758 and published his observations in *A Synopsis of the Astronomy of Comets* in 1705. He would not live to see the comet's return.

In 1949, Halley's Comet became the namesake, and name-play, of the band Bill Haley & His Comets, famous for their hit, 'Rock Around the Clock', a defining moment in rock 'n' roll which was also outwardly perceived as Haley's personal perihelion.

Haley struggled with fame, with chronic alcoholism and the effects these had on his marriages and ability to support his children. Haley would never see the comet whose coattails he had flown on as it swung around the Sun. Halley's Comet would next return to Earth's skies in 1986, five years after Bill Haley died, aged 55.

This rather cruel twist of fate was loosened, just a little, in 2006 when the International Astronomical Union announced the naming of an asteroid to mark the 25th anniversary of Bill Haley's death. Asteroid 79896 Billhaley now floats through the rhythmic infinity. There is some poetic justice in a man once perceived as a rock 'n' roll pioneer having a giant space rock named after him.

As for Mark Twain, his orbit of the Sun mirrored the path of the comet. 'It is coming again next year,' Twain said in 1909 of Halley's Comet, 'and I expect to go out with it.' He died on 21 April 1910, just as he had been born, during the month of the comet's perihelion. Halley's Comet will next return to our skies in 2061, reaching its perihelion on 28 July.

# Comet Vintages

In 1858, in a wooded Dorset parish called Yellowham, or 'Yell'ham' as the locals pronounced it, an 18-year-old writer gaped at the sky as a looming, curved comet rambled over the bulbous, arable hills, golden under the summer sun. There, a second daytime star had invited fear in the backward, and awe in the more enlightened of the parish. Some even dared to wonder if it might be a portent of a good year.

The young writer was Thomas Hardy, and he was suitably inspired by the comet's appearance, and the distant promise of its reappearance, to pen a poem.

## The Comet at Yell'ham

I

*It bends far over Yell'ham Plain,*
*And we, from Yell'ham Height,*
*Stand and regard its fiery train,*
*So soon to swim from sight.*

II

*It will return long years hence, when*
*As now its strange swift shine*
*Will fall on Yell'ham; but not then*
*On that sweet form of thine.*

In a world that was now governed by the scientific insight of Copernicus, Kepler, Newton and Halley, to name but a few, comets such as this became known in the Victorian era as 'flowers of the sky' and were largely perceived as being exciting rather than terrifying, inspiring outpourings of creativity in both the arts and sciences. And this comet would be a special one.

Now known as Donati's Comet, this long-period comet is named after Giovanni Battista Donati, the Italian astronomer who first observed it. Donati's Comet was the first comet to be captured in a photograph, and in some images, the bright, wide-tailed comet appears like an ectoplasmic bloom across a sepia space. That bright,

wide tail established it as the most brilliant comet of the 19th century, which inspired the hopes of many that this might be a very good year indeed – especially for those working in certain vineyards.

It may have been coincidence, perhaps there was just something in the air, but since the Great Comet of 1811 and the remarkably high-quality harvests experienced by vineyards that year, wines produced during comet years became extremely covetable – and exceptional. In 2011, a bottle of 1811 Château d'Yquem, a sweet Sauterne, was sold for £75,000.

Similarly inspired by the 1858 comet, Charles Dickens referenced the notion of good wines coinciding with the arrival of comets in *The Uncommercial Traveller* (1860–61) and these wines became known as 'comet vintages'. This was also referenced by Arthur Conan Doyle in Sherlock Holmes's 'The Adventure of the Stockbroker's Clerk' (1893).

For those hoping for another stellar vintage, Donati's Comet was a blessing from on high. 1858 did turn out to be a very good year: it became known as 'a year dear to all Claret lovers'. Should you ever be lucky enough to be granted a sip of that particular comet vintage, savour it. Donati's Comet won't be seen again until the year 3600.

# Love's Arrow

Early in my relationship with my husband, we borrowed a friend's ridiculous leopard-print tent and camped in Elan Valley, Mid Wales, in early September. Elan Valley is a Dark Zone, where the night sky is as clear as glass, and the stars of the Milky Way coalesce like a breath upon the pane. We lay out on the grass, deep into the night, roving the skies with binoculars bought from a local charity shop, removing them at intervals to watch and point to one another as a meteor from the minor Perseids tore through the sky.

Lovers – requited and unrequited – have long looked for shooting stars as evidence of their affection, perhaps none more so than Orpheus, the poet and musician of

Greek myth. Orpheus didn't just play the lyre, he created music with such magnetism that both mortals and deities were drawn to the sound and even the trees swayed in time. From these trees a beautiful woodland nymph called Eurydice emerged, lured by his music. As soon as their eyes met, it was love for both Orpheus and Eurydice, and they were quickly married, hoping to live a long life together: to sit in chairs by the window as the white, Greek sunlight poured in, and Orpheus played love songs to his wild wife.

But the Fates had other ideas. Stepping out of their home one day, Eurydice was bitten by a snake and died. Unable to live without her, refusing to, Orpheus scoured the Earth for her spirit, descending into the Underworld. There he played every requiem, elegy and lament. His instrument sang of such sorrow that the gods relented. Orpheus would be permitted to retrieve Eurydice and return her to the land of the living, under one condition: that Orpheus left the Underworld first and did not look back to check on Eurydice.

He looked back.

Orpheus spent the rest of his life alone, playing songs of lament dedicated to his wife. When he died and was finally reunited with his love, the gods hung his lyre in the night sky, in memory of his music and his eternal devotion to Eurydice. So goes the legend of the Lyrid meteor shower, which occurs every April as the Earth transits through the debris of the Comet Thatcher, a long-period comet that orbits the Sun every 415.5 years. The Lyrids have been observed for over 2,700 years, making them the oldest known meteorite shower.

Like seeing a shooting star in the skies of the 21st century, the arrival of love to lovers can seem quasi-miraculous. In the sea of stars, that there should be one that burns bright – however briefly – for you, and that they might grace your life with their brilliance is surely, nearly impossible.

In the little garden of the thatched cottage where my husband and I now live, we are lucky to be remote enough to see the Milky Way, past which shooting stars burn through the granular black. Watching the night come down in a nearby field, waiting for the stars, I watched a swarm of swallows. They were a storm of black arrows. Out of the maelstrom, out of the seemingly infinite possibilities, a mating pair found one another and broke away to perform their aerial dance in the empty sky with one another. It felt like watching two comets collide.

# PART TWO

# When the Stars are Smelted:

## *Utilising Heaven's Rocky Bodies*

# Cosmic Energy

When I drink my coffee and write my diary in the morning, I sit and stare out across the field at a high, white wind turbine that towers above the patient oaks on the horizon. In as few as two swift turns the turbine has produced enough energy to power my home for a day. It is considered to be 'clean' energy. The only problem is that wind turbines, and other 'clean' machines like electric vehicles, are powered by permanent magnets that use elements – such as neodymium (Nd), praseodymium (Pr), dysprosium (Dy) and terbium (Tb) – that require large quantities of energy and cause a vast amount of environmental destruction to be sourced from the Earth. But, if we can find a way for the wind turbine

I watch every morning to be powered by an element that only occurs naturally in meteorites*, there are hopes that, one day, the initial environmental cost of clean energy will be nothing more than an uncomfortable memory.

Before it was discovered that the alloy tetrataenite (FeNi) existed in the cosmos, in 1962, the future Nobel Prize winning physicist Louis Néel and his co-workers synthesised tetrataenite for the first time using neutron irradiation – a process that is not suitable for mass production. Almost 20 years later, tetrataenite was discovered in meteorites, leading to the theory that this iron-nickel alloy was formed organically in the cosmic blast furnace of an asteroid, through the process of iron and nickel atoms cooling over millions of years. Thus, the race to find a sustainable way to produce this element in a lab began. At the time, it was believed that a process of extremes was required – either cooling over millions of years, or neutron irradiation.

Cambridge University's Department of Materials Science and Metallurgy led the research with the Austrian Academy of Sciences, and they appear to have found the missing link: phosphorous, an element that is also present in meteorites. It is still early days for their research and the

* One rock containing tetrataenite believed not to be of extraterrestrial origin was found in 2015 in the Indo-Myanmar Ranges of north-east India.

subject of much debate, but it is thought that phosphorous encourages the nickel and iron atoms to move more rapidly, the result being that tetrataenite can be formed in a few seconds, rather than a few million years.

If tetrataenite becomes readily available as a resource to power clean energy, the costs of production will be reduced, the environmental impact will be almost negated and our reliance on fossil fuels will decrease to the extent that our energy may become almost entirely clean – the energy used to produce the coffee I drink will be as limitless and abundant as the sky the turbine cuts through. Perhaps our next challenge will be discernment: becoming more judicious when considering what is truly worthy of that energy, and whether all of the devices powered by clean energy benefit all life on Earth.

Progressive – futuristic, even – as this development in clean energy may seem, it is following a millennia-old framework that humankind has designed using materials from shooting stars. Since our beginning, we have found practical uses for these fragments that fall from the sky and the heavy metals that they contain. Historically, they have been our companions – portals, even – into other, imaginary realms; they have sustained isolated populations and provided the munitions for basic tools; they have been the basis of superstitions through which we might hope to cure ourselves of ailments; they have created wedding bands; been at the root of smuggling rings; and they have unlocked some of the best-kept secrets of existence. A shooting star is so much more than something to admire. In its most advanced applications it can tell us when, and from what, our Solar System originated; and in its most primitive it provides the materials for humankind to survive hostile environments. Cultures, and even entire civilisations, have prospered thanks to the materials that have rained down on our little blue planet. And as these materials settled like dust on Earth's surface, our ancestors asked not 'Can we use this?' but 'How can we use this?'

# Buried Blades

The desert conceals a blade comprised of the tattooing of crystalline structures, not of this Earth. They appear like a prisoner's tally lines on a cell wall and are known as lamellae. But rather than marking the days, weeks, months or years of incarceration, they indicate small plates or flakes of kamacite or taenite (iron and nickel alloys). The pattern they produce cannot be forged by a human hand. It is known as the Widmanstätten pattern: a structure only found in meteoric iron, formed during the slow intergrowth and cooling of iron-nickel within an asteroid over millions of years, out in the smithy of space.

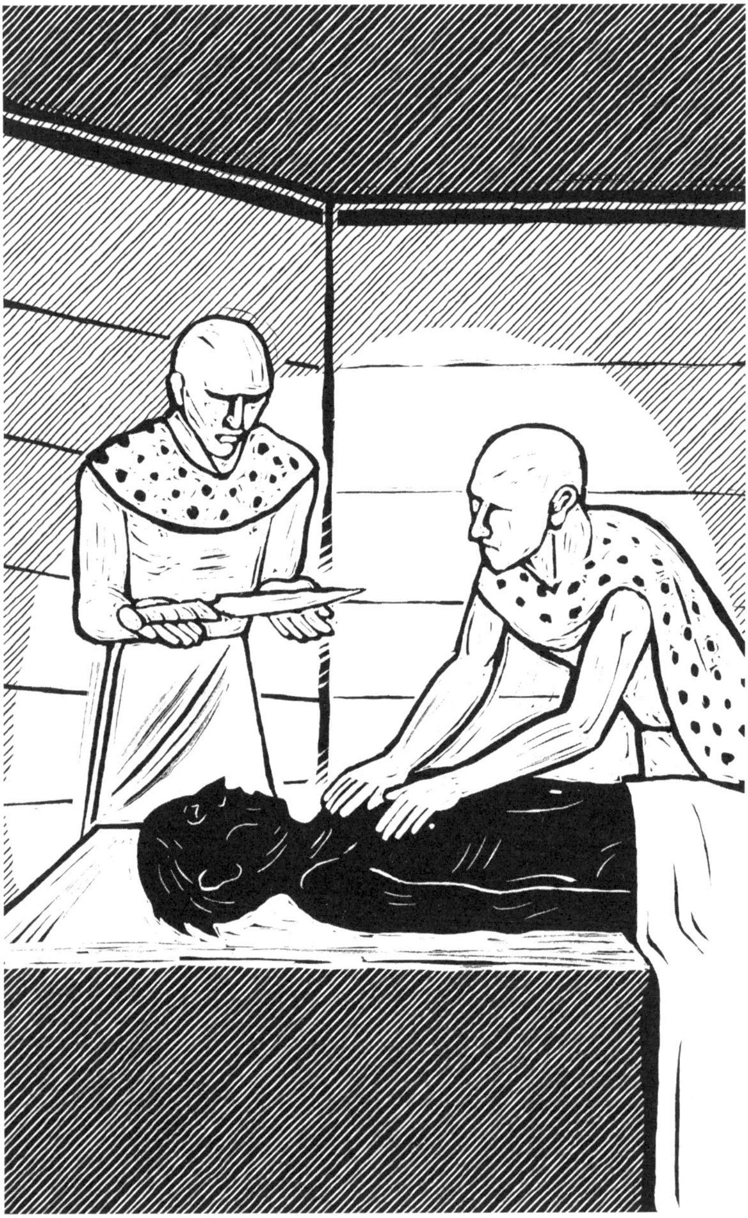

There is no definitive evidence that Ancient Egyptians were aware that iron-rich meteors fell from Heaven to Earth, but the ritual and ceremonial use of their materials indicates that, in all likelihood, they had a very good understanding of where these metals came from. That they were aware the iron for their most prized blades came from space is demonstrated by meteoric iron blades being referred to in temple inventories as 'stars'. These origins would have given meteorites both powerful ritual magic in certain settings, and the use of their alloys would have imbued any otherwise quotidian instrument with a sense of otherworldliness, even divinity.

In one of the most beautiful examples of ritual use, a meteorite's metal was crafted into a ceremonial dagger with a hilt of fine gold cloisonné, at the end of which is cool, clear rock crystal. Believed to be a gift for the pharaoh Amenhotep III (1401–1353 BCE), the blade's sheath is also made of gold, decorated with an intricate foliate design.

Amenhotep III was the making of the Ancient Egypt we know today. As demonstration of his success as a pharaoh, his political dominance and strategic alliances, he had many wives, one of which is thought to have been the daughter of Tushratta. Tushratta was king of the Mitanni – an empire stretching from modern-day Syria to western

Turkey, where the ceremonial use of meteoric iron was not uncommon and their metalwork was highly advanced in both production and decoration. Diplomatic documents from the time called the Amarna letters mention the gifting of iron objects from Tushratta to mark this union, and it is generally thought that this dagger was among those gifts, with iron – particularly meteoric iron – worth more than gold in Ancient Egypt. But the dagger wasn't discovered in Amenhotep's tomb, as one might have expected; it was found in the tomb of his grandson, the golden boy, Tutankhamun.

The Egyptian Book of the Dead records that iron knives were used during the embalming process for the 'mouth opening ceremony'. The most important rite of the embalming ritual, it was performed by a priest in leopard-skin robes. In this ritual, iron knives were used to open the deceased's mouth, ensuring the soul could breathe in the afterlife and that their spirit would be reanimated.

It is believed that the meteoric blade of Amenhotep's dagger was used during the ritual of Tutankhamun's embalming, at the age of 18 or 19 – around 3,250 years before the dagger was recovered. Looking at the blade from this perspective, the series of interwoven scratches and dashes of the Widmanstätten pattern marked on the dagger's blade appear as if some celestial hand had pre-empted its time underground, tallying the years it knew the dagger would spend in the dark cell of the funeral tomb, aeons before the iron ever fell to Earth.

# Prescribing Stars

In a room of thick stone, limewashed a grey-white, a man with a large tumour on his neck looks trustingly at his physician. The man is desperate; not only has his self-confidence dwindled in the years this lump has grown, it has also begun to affect his sleep and his ability to perform his job as a stonemason. With long hours required with his neck in awkward positions, the tumour has started to impede the angles that he needs to get himself into if he is to perform his job to the standard he prides himself in. His wife has also begun to worry that it might represent something more sinister. The stonemason wants rid of this lump.

The patient and his physician stand before the open fire, over which the stonemason's wife has left a stew to simmer. The medical practitioner hands the stonemason a parcel wrapped in muslin cloth, and sternly murmurs his instructions for how these remedies must be applied: specificity is paramount. The stonemason nods gravely, this might be a matter of life or death; with his livelihood in the balance, it is certainly a matter of survival.

The stonemason dutifully follows the directions of his physician, first cutting a length of rope. He then uses this rope to secure one side of a verbena root to the tumour on his neck, ensuring the other side of the root is slung over the smoking fire – the smoke is very important. The stonemason must stay like this, leaning over the fire with the root attached to his neck, for several hours – preferably days – if he is to cure his tumour. And there, in the midst of hope, before the slow drip of disappointment begins, we shall leave the stonemason.

In ancient Gaul, as was often the case in Celtic Britain, magico-medical practices were the solution to ailments before the modern concept of pathology had developed. With chants and spells, homeopathic magic was used to venerate allies or destroy enemies through the use of their image, or cure ailments through charms, herbal remedies, incantations and amulets. This type of sympathetic magic swung between extremes, taking the form of human sacrifice, or, as we've just heard, using a verbena root and the smoke of a fire to cure a tumour.

Magico-medical practices are believed to be the origin of the word 'Abracadabra', which was first found in a medical text. Written out nine times in a triangular form and worn as an amulet, it is believed to have been a cure for, or used to ward off fever:

*abracadabra*
*abracadabr*
*abracadab*
*abracada*
*abracad*
*abraca*
*abrac*
*abra*
*abr*
*ab*
*a*

These practices focused on invoking the power of an invisible, intangible force through the ritual performance of certain rites while employing physical entities, like a verbena root – or a shooting star.

Marcellus of Bordeaux, also known as Marcellus Empiricus, was a medical writer and possibly the court physician to the late-period Roman Emperor Theodosius I around 394 CE. Marcellus compiled a compendium of folk remedies, medicines and pharmacological preparations in his medical text *De medicamentis liber,* which organised remedies from the head down to the feet. In a remedy for acne, Marcellus advises those afflicted to 'dab their pimples' with a cloth while watching a falling star, asserting that the spots would 'fall from the body', reflecting the course of the star. However, if the sufferer does not follow the instructions to the letter and wipes their pimples with their bare hand instead of a cloth, they will soon find the spots have been transferred to the offending palm.

Although I haven't found this method effective myself, under the supervision of Marcellus and his followers it was best not to appear as a thankless or sceptical sufferer – he also included advice on how best to avenge ungrateful patients. In the case of our poor stonemason, should the medical practitioner deem him unappreciative, the verbena root would be retrieved from the smoke and rehydrated, with the aim of worsening the tumour further. Those suffering from pimples, holding a cloth to their face while desperately searching for a shooting star, might also be advised to wish for a miracle when they finally see one.

# The Iron Mountain

A tusk that has visited the farthest fathoms of the ocean is united with metals that have seen the very depths of space. Known to the Inughuit – the Inuit people of the northernmost area of Greenland – as Innaanganeq, meaning 'great iron', the Cape York meteorite is one of the largest iron meteorites to have fallen to Earth, with the heaviest of its eight fragments weighing 34 tons. For the Inughuit, this meteorite provided the primary source of iron to make survival tools long before European trade arrived. In this instance, the Inughuit transformed a narwhal tusk into a lance using cold-forged iron teeth from the meteorite, which fell to Earth around 10,000 years ago.

Innaanganeq, like other iron meteorites, was once the metal core of an asteroid that is thought to have been destroyed in a catastrophic collision with another asteroid around 650 million years ago. Innaanganeq is a medium octahedrite (an iron meteorite that contains nickel) and is almost as old as the Sun, dating from around 4.5 billion years ago. It entered Earth's atmosphere at around 12–16 miles per second (20–25km/s), which caused it to fragment. As this fragmentation took place around 60 miles (100km) above the Earth's surface, the strewn field of this meteoroid was vast.

The Inughuit had known of the meteorite's presence in north-west Greenland for centuries, probably millennia. There was a legend about its three largest fragments, which they called the Woman, her Tent and her Dog. These had been hurled from Heaven by Torngarsuk – a deity who sometimes appeared as a bear, sometimes as a one-armed man'. One of the higher-ranking gods in the Inuit pantheon, Torngarsuk is master of whales and seals, and god of the sea, death and the Underworld. The eight largest fragments of Innaanganeq weigh a total of around 58 tons; the largest fragment, the Tent, weighs 31 tons alone.

The Inughuit, and the Dorset culture (Paleo-Eskimo) before them, had made use of the meteorite's fragments for metalworking since the Stone Age and, due to a limited amount of driftwood for fires washing up on their shores,

transitioned to the Iron Age without the capacity for smelting. Theirs was a more difficult method of cold forging: breaking off, hammering and shaping the metal with basalt hammerstones weighing up to 90lb (40kg) to create lances and harpoons. In a region where survival depended on sharp tools for hunting and skinning animals for both food and warmth, the source of iron provided by Innaanganeq was a lifeline.

Subsequent analysis demonstrates a trading culture between the Thule people (ancestors of present-day Inuit), the Norse and the North American mainland, and shows that the Thule supplied Canada and North America with metal, the meteoric iron. The Inughuit had exclusive access to this meteorite for well over a thousand years until men with adventure on their minds began to arrive. But first, an Inuit had to leave.

An Inuit called Sakæus harboured desires for horizons that stretched beyond his vast white and black island. In 1816, he stowed away on a visiting Scottish whaling ship and arrived in Leith, Scotland, later that year, where he quickly learned English and demonstrated his prowess as a kayaker and harpoon thrower to large crowds. He also became a talented painter. With these skills, combined with a detailed knowledge of the geography of Greenland, Sakæus would become a vital crewmember for European explorers.

Following the Napoleonic Wars at the beginning of the 19th century, Britain had a surplus of ships and decided to employ them in the search of the North West Passage, a route that had been sought by Europeans since Columbus in 1492 and one that would drastically shorten the trade route to Asia. To this end, a man called Captain John Ross and his crew, which included Sakæus, reached Melville Bay on Greenland on 18 August 1818.

While there, Ross observed the handiwork of Sakæus's harpoon heads and became curious as to the source of the metal. While Sakæus almost certainly knew that the metal of his harpoon heads was made from meteoric iron, he was not familiar with the exact source. Sakæus overcame the lingual and cultural barriers that existed between the Inuit and the Inghuit of the High Arctic and, over a mere ten-day period, he learned that the meteorites' location was at a place called Sowallick, meaning 'iron mountain'.

The expedition did not attempt to reach the impact site, but Ross did purchase meteoric iron knives from the Inghuit, which he agreed were of extraterrestrial origin (something that had only been accepted by rational science around 15 years earlier and was still disputed). That Sakæus and Ross had been told the correct location of Innaanganeq would not be confirmed for another 76 years, when an American explorer named Robert Peary arrived in Greenland.

*

The Inghuit were naturally protective of their meteoric iron at first. But over 12 years of expeditions, trying and failing to reach the North Pole, Peary, a former civil engineer in the US Navy, built up a relationship with them. In 1894, on an expedition to northernmost Greenland, two Inuit named Panikpah and Tallekoteah took Peary to the 'iron mountain', and there they showed him three of the largest fragments.

Peary then took it upon himself to remove the Woman, the Dog and, finally, the Tent, which his daughter renamed Ahnighito, after her middle name. Peary was aware of meteorites' importance to the Inghuit in both a practical and spiritual sense, but he was running out of funding to get to the North Pole, and needed something to show for his expeditions. It would take three years and the building of a railway, but eventually Peary took all three pieces back to New York, together with six Inuit, on the understanding they would be given tools in return for consenting to being studied before being returned to their homeland.

In 1909, Peary claimed he had finally reached the North Pole, thereby achieving the fame and acclaim he had always desired. Later analysis of his records in fact suggest he never made it, falling short by several miles due to navigational errors. This pyrrhic expedition was made possible thanks to the sale of the three largest meteorite fragments, finalised that year by Peary's wife, Jo, who sold them to the American Museum of Natural History for $40,000 (equivalent to around $643,000 today). Over the years that followed, the meteorite's remaining large sections continued to be taken from Greenland.

Four of the six Inuit that Peary brought back to America with him, under the false promise of their return within a year, died shortly after arrival, having been studied and paraded like the meteorite. Minik Wallace survived that first year, but his father had been among those who had died. A burial was staged to deceive Minik into believing his father had been lain to rest. In actuality, his father's skeleton was put on display at the American Museum of Natural History. It was kept there until 1993, 75 years after Minik's death, when his father's skeleton was eventually repatriated.

The same cannot be said for the meteorite that helped keep a whole community of people alive for millennia, or any of the funds that Peary's family received for its fragments. Having been sufficiently analysed long ago, the meteorite is now kept thousands of miles away from its original point of impact. The largest fragment, the Tent, is so heavy that its supports in the American Museum of Natural History have been tunnelled into the bedrock beneath the building – as if there is the concern that, one day, the iron mountain might decide to wander home.

# Dark Markets

> *Alleged meteorite smuggler Walter Horst was granted bail of N$10,000 yesterday, with stringent conditions attached, after close to a month in police custody. Horst (58) has to report to the police three times a day, is under effective house arrest between 19h00 and 07h00, may not leave the Windhoek district without permission from the police, and has to surrender his travel documents to the police.*

Owing to the development of internet-based, unregulated trade, a boom in commercial demand, and the discovery of a number of isolated and unprotected strewn fields in Africa and the Arabian Peninsula, the early 2000s was a time of burgeoning meteorite smuggling. And this contemporaneous news report from 2002 details the restrictions placed on suspected smuggler Walter Horst. Horst, an Austrian-born Namibian citizen, had already been arrested previously for allegedly attempting to smuggle fragments from the same meteorite. In 2001, acting on a tip, police intercepted a transaction in which Horst was allegedly paying a deposit of N$10,000 of N$30,000 to four Namibian men for a 650lb (296kg) fragment of the Gibeon meteorite.

The Gibeon meteorite fell near what is now the village of Gibeon in Namibia between 13,000 and 30,000 years ago. Its vast strewn field of iron rain – around 240 miles (386km) long and 75 miles (121km) wide, one of the largest on Earth – was distributed across the remote regions of Hardap, which spans the Namib Desert. As with the Inghuit communities in Greenland, the ancient Nama peoples had been using the meteorite for millennia as a resource for tools – such as arrowheads, spearheads and even rudimentary bullets.

At the time, the authorities were suspicious of mineral dealers, and Horst was known to make (what may have been completely innocent) monthly trips to southern Namibia, the region where most of the Gibeon meteorite fragments were found. During a subsequent bail application, Horst testified that he had been collecting 'minerals' in the south of Namibia for a decade as part of his business, and insisted these activities were not suspicious. During the search of Horst's house police found an illegal rhino horn, for which he was also charged, but not the stolen meteorite.

Following the Gibeon meteorite's 'official' discovery in 1836 by the English captain, J E Alexander, a collection of 33 of the largest samples (weighing between 195 and 506 kilograms) was undertaken in 1911. These samples were originally displayed in a pile at Namibian capital Windhoek's Zoo Park; but by 1975, the meteorites had been designated a National Monument and a plan for their formal public display was initiated. To ensure their safety during the construction of the display, the pieces of meteorite were moved and stored at a fortress in Windhoek – this is where the first two of the fragments were stolen. The display continued as planned and was erected in the capital's Post Street Mall, with two empty pillars to represent the stolen fragments. Since then, two more fragments have been purloined.

Although jewellery companies tend to maintain that pieces made from meteorite are covetable for their rarity, it is actually for their novelty. Transported in the form of falling bodies of various sizes, metal alloys from space are more abundant on Earth than gold. It is the magnetism – both metaphorical and often literal – of an object that has fallen from realms beyond our reach that gives the otherwise commonplace elements of nickel and iron, for instance, their allure. That, and their desirable feature of the Widmanstätten pattern: the interleaving lines that are like a watermark, impossible to forge, not visible to the naked eye – unless it has been treated with a weak acid solution, and are unique to each meteorite. This novelty has generated a healthy black market trade of meteorites that are regarded as national treasures in their countries of impact.

'It was easier than I expected,' an American said of smuggling meteorites out of Algeria in the early 2000s. 'I thought I would be lugging kilos of meteorites out of the country by plane, but we found a shipping company at the port that was happy to ship them out without asking any questions.'

Meanwhile, Horst just couldn't catch a break. In 2002, he was arrested again, this time accused of lying to a shipping company about the contents of a crate he was transporting. It is believed that, in order to disguise the Gibeon meteorites he was allegedly attempting to send to South Africa, Horst told the shipping company that the crate contained 'an engine'. But if there had been anything other than an engine in the crate, the police were too late.

The two cases against Horst were consolidated, but the charges against him and the four other men were dropped in 2004 after procedural delays. The stolen fragments of Gibeon meteorite have never been recovered. But this case brought to attention the illegal trade of meteorite smuggling in the early 2000s, with museums such as the National History Museum in London and the Smithsonian in New York now refusing to accept meteorites unless a proof of origin can be supplied.

Following the case against Walter Horst, the Namibian government passed the National Heritage Act, making it illegal to export any meteorite from the country. It remains illegal to possess newly retrieved fragments of the Gibeon meteorite, but those retrieved before the law was passed are still in circulation, still turn up in jewellery and occasionally, as one man boasts online about his piece, turn up in house clearances.

Of the original 33 fragments of Gibeon meteorite, 29 are on display in Windhoek. The others may exist in their entirety in the homes of collectors, or they may have been melted down, or fragmented and set in some of the many meteorite rings worn on the held hands of the recently betrothed.

# The Šaman's Amulet

Firelight glowed upon the ridges and pooled in the hollows of the walls. In the recess of a limestone cave, in shrinking shadows and changing light, shapes and patterns painted by a seer came to life – animated, they played out an otherworldly scene, designed to aid his metaphysical enquiry. As he stared, unblinking, at the evolving vision before him, the master rocked himself until he crossed the veil, falling into the imaginary world before him. He became one of those shifting shapes in the amber light, and drifted into his trance, clutching a sparkling star.

The object held in his aging hand aided his travel through the walls of the cave, through the boundary of reality as he knew it, and into a place that was constant: a spirit world where only truth was spoken, albeit in riddles; a realm where he could find answers, reasons, cures, curses and prophecies. To someone without these gifts, his might be perceived a lonely life, largely removed from the rest of the tribe. But the shaman laughs, because he has his spirits, and he has his glistening tools.

Typically associated with depictions of wild animals, ancient cave art also featured shooting stars and spiritual masters such as the man described above. The abstract patterns and images occasionally found adorning the walls of caves are now believed to be visions of the shamanic cosmos. These archetypal images are hard-wired into the human nervous system and are known as 'phosphenes', 'entoptic' images or 'form constants' – phenomena experienced in light deprivation, or images that originate within the eye itself: scintillating, recurring geometric patterns. They can be seen as a very primitive form of language, a means for someone of power – a shaman – to communicate with their followers, and were likely induced by trance-like states, sensory deprivation, dreams and occasionally the ingestion of hallucinogens like psilocybin.

Shamanism is a phenomenon that was first recognised among the Evenki peoples of Siberia, from whose language the word *šaman* was appropriated as a generic label for similar practitioners among neighbouring or related peoples. The etymology of the word 'shaman' is disputed, but what is known is that its recognised form originated in Russia and these ancient spiritual masters held the same positions we associate with shamanic practitioners today – the ability to heal or wound, to commune with the spirits and the divine, and communicate with and through the natural world.

Prehistoric shamans have been connected to meteors through depictions in cave art across the world, from Morocco, to Algeria, to Utah. These include the representation of a human with a be-ferned or antlered head, around their head birds swirl, and, flying past a writhing snake that the figure holds in their hand, are four meteors. When those meteors fell to earth, meteorites were also coveted in their physical form, as amulets, by our ancient ancestors. It is believed that they were seen as ways in which a shaman might find inroads to the otherworld, or to ward off and interrupt the evil from those worlds through the use of ritual magic, or apotropaic rites. And it was an amulet found in the humble dwelling of a man believed to be an ancient Slavic shaman, found in modern-day Poland, that would offer up the earliest known evidence of a shaman's connection with meteors.

In Poland in 2013, archaeologists near Szczecin, northwest Poland, discovered what they believe to be the remnants of the home of a prehistoric shaman from the Early Holocene, roughly 11,700 to 7,000 years ago. Among the relics associated with a spiritual and magical culture they uncovered were an engraved bone spear tip; a stick made from an antler engraved with geometric, triangular patterns; and a wooden censer for burning incense-like substances such as resin. There was also a small rock that, when lifted, was of surprising weight. This was a pyrite (iron sulphide) meteorite, known as fool's gold.

Several large meteor showers are known from prehistory Poland. The closest to the shamanic meteor find was the Przełazy fall in western Poland, thought to have occurred around 10,000 years ago. However, this fall largely consisted of chondrite (stony iron and nickel) meteorites. So, exactly where this prehistoric shaman acquired his meteoric fool's gold is a bit of a mystery.

Its uses are easier to deduce. The ancient Slavs saw the world as a conclave of forces, manifest in both the animate and inanimate; and intrinsically connected to power, to perceived magic, were meteorites. Being not of this Earth, and sparkling like a star, the meteorite may have been regarded as holding particularly potent supernatural powers in its own right, thereby enhancing the shaman's supposed divining abilities. Things that originated from *up there*, from divine, powerful forces, where the spirit's after-life plays out, were imbued with that same penetrating, ancestral power. With this passport from the treasure chest of our galaxy, the shaman could live between worlds.

# The Winter Years

Presolar grains, or stardust, were once part of the lifecycle of a star; they were part of its beginning, and comprised its deathbed. When a star burns out after millions, or billions, of years, it ejects the condensed gases (now in the form of particles) that formed it back out into space. These solid interstellar particles may yet form new stars, planets, moons or a meteorite.

If stardust – from a red giant (stars in their winter years, like our sun); supernovas; novas; Wolf-Rayat stars (thought to be the massive precursors to supernovae); and even 'born again stars' (where a white dwarf experiences a new lease of life in the form of a bright, hot helium flash) – is sufficiently synthesised, it will create a primitive meteorite. If that primitive meteorite then finds its way to Earth, it will contain presolar grains: the secrets of the genesis of our cosmos.

The theory of presolar grains was first proposed in the 1960s, during the study of the isotopic formation of the Universe. Two American scientists, John Reynolds and Donald Clayton, discovered that primitive meteorites were rich in the element neon (Ne), a component that likely meant meteorites contained residue of stellar ejecta that had survived the formation of our Solar System. Clayton predicted that several types of presolar grains would be found in meteorites from various star formations. Few supported his theory at the time, and it was only later confirmed because of the timely arrival of a meteorite.

On 28 September 1969, a medium carbonaceous chondrite meteorite landed near Melbourne, Australia. Now known as the Murchison meteorite, it fragmented, and one part of the Murchison meteorite tore through a barn roof and landed in a bed of hay. Throughout its descent through the atmosphere, and the barn roof, the meteorite remained relatively unaltered, its secrets concealed.

The meteorite was retrieved and advancements in microanalysis developed over the intervening years. Further study of the Murchison meteorite by Edward Anders and his team at the University of Chicago in

1987 confirmed what Reynolds and Clayton had long suspected. Presolar grains are the only physical example we have of stars.

They present us with a microscopic porthole into the Universe's past. Through their analysis we are offered impressions of how stars live and die, how they synthesise, and we are even offered glimpses as to how our Solar System was formed, 4.5 billion years ago. Before the discovery of presolar grains, all our knowledge about stars and their lifecycles came from observation and the electromagnetic radiation they emit.

Using diamond knives, analysis of the Murchison meteorite continues to this day. It is one of the most studied meteorites and, in January 2020, further analysis determined that the particles it held were around 7 billion years old, approximately 2.5 billion years older than our Solar System.

Since 2010, efforts have been made to primitive asteroid samples from space rather than waiting for them to be delivered to Earth as rare primitive meteorites. This is because, not only do presolar grains hold information about the building blocks of the Universe in the form of amino acids (some of which are not present on Earth), they also provide insight into the building blocks for life itself. This fundamental understanding has been achieved because of falling stars, which have brought dust from furthest reaches of space into our remit, thereby offering a fingerprint from the invisible hand of creation.

# PART THREE

# When the Star Falls to the Left:

## *Belief, Lore & Superstition*

# What Light is That?

## Global

Many are the nights you will have stared up at the dark infinity, at the Moon and its chorus of stars – at the static magic of it all, unmoving to the casual observer you are – and longed to see something more. Something fast. Something bright. Something fleeting. A little firework in celebration of nothing but your own capacity to witness it.

> *'Who's there? What light is that? Wherefore com'st thou?'*
> *'To comfort you, and bring you joyful news.'*

Poet and playwright Christopher Marlowe wrote these words 12 years after the Great Comet of 1577 blazed through Europe's night sky. The Great Comet was a shooting star that made a singular contribution to our understanding of the Solar System. Danish astronomer and alchemist, Tycho Brahe, made accurate measurements of the comet's trajectory during its three-month journey across the sky. Collating and comparing his own measurements with those of other astronomers in Europe, Brahe was able to demonstrate that the previous idea that comets were heavenly phantoms, confined to the realms beneath our Moon, was a fantasy. Brahe's calculations determined that the comet travelled through planes that existed beyond the Moon – at least four times further away than our silver satellite – thus proving that comets were not the 'windy exhalations' of Earth's atmosphere, as had been proposed over 2,000 years ago by Aristotle in ancient Greece, a civilisation that hereto had formed the bedrock of Western understanding of the cosmos.

Brahe, like many scholars of the Enlightenment rediscovering classical texts, had read (and corrected) the Ancient Greek astronomer Ptolemy, who encouraged the perception that shooting stars were auspicious. Ptolemy wrote that to see a shooting star was evidence of gaps between the mortal and divine cosmic spheres, extant when the gods peer down onto our little lives and listen to our wishes. This is likely the origin of our modern inclination to 'wish upon a falling star', and their connotations with the divine, the miraculous and with luck.

Today, it is the scarcity of the shooting star rather than its mystery that inclines us towards magical thinking. We might not make a wish, but with light pollution worldwide reducing our ability to see these phenomena, simply to be in the right place at the right time will likely predispose us to feel chosen, special, and to see the star as Ptolemy did: as a good omen, a nod to hope.

Only rarely do these portents remain in the sky for longer than the split second it takes to make a wish, or burn bright enough to be seen in cities. But in October 2024, amateur star-gazers in the southern hemisphere, and soon after the northern, were blessed with the comet Tsuchinshan-ATLAS, a bright periodical comet that had travelled from the Oort Cloud – a stony kingdom at the very edge of our Solar System, and the birthplace of long-period comets.

As the comet travelled through the endless skies of the Atacama Desert in Chile, above the strange sandstone structures of Utah, over the tor on Glastonbury Hill, hundreds of thousands of ordinary people made something akin to a pilgrimage at twilight to witness a light burning in the western skies, knowing that it would never be seen again by the same eyes.

Despite centuries of rational knowledge about this celestial event, the comet still inspired feelings of awe, of the miraculous. It will not return for another 80,000 years. To many, that existence had conspired for us to be alive at just that moment in time was, indeed, miraculous. Unless you lived in mid-Devon, that is, where the sky clouded to the west at twilight every single evening, without fail, for those days in October, completely obscuring the comet's line of flight. 'We'll see it next time,' my husband reassured me.

In *The Comet*, published in 1869, author Haywood Williams Guion lamented that the human condition always perceived the miraculous as something distant, like a shooting star. 'The human mind,' he wrote, 'unimpressed with familiarity, has ever fondly conceived that the rich prize of hope must reside amidst the enchanting scenes of distance, and not in the simple and plain fields of home.' Our 'plain fields of home' are made of the same stuff as shooting stars, and the visitation of that stuff from the black urges us to throw our imaginations beyond the immediate, to the origins of mystery – that is the prize. Not one or the other, but both simultaneously.

In 2011, the peace of the aptly named Comette family was shattered when a meteorite smashed through the roof of their home on the outskirts of Paris. The egg-sized object was valued at £75,000, but the Comettes said selling it was out of the question. 'A piece of the history of space of which we know nothing, but which is fascinating, has fallen on us,' Mrs Comette, told *Le Parisien* newspaper. 'It's like a fairytale.'

The Comette family's view of the 'shooting star as enigma' is just one of myriad ways these sparks have kindled the imagination of humanity. From such stuff we have forged creation myths, domestic superstitions, sacrificial rituals and entire belief systems shared down the caravan of generations through oral lore, carvings in stone, the inking of animal hides, and the continuation of sacred rites. Though ephemeral and fleeting, shooting stars seem to speak directly to the human soul and, in doing so, they speak of the eternal, of the forming of planets. They speak of all that is pre-human, pre-life as we know it. And these 'stars' that seemingly appear out of nowhere encourage us to question what is beyond the immediately visible. They prompt us to wonder at what might be beyond the horizon, where the shooting star is born and ultimately vanishes, and to hope.

# A Bull with Heaven on His Back

## North America

Shooting stars were portents of war, the revealing of a secret location, travelling spirits of shamans and heroes journeying to the afterlife, or signs that a great chief had just died for the Blackfoot, the Seneca and the Cheyenne. The Native Americans of North America have had as many beliefs for shooting stars as they had unique cultures – they were not simply binary, interpreted as 'good' or 'bad'.

The Skidi Pawnee of what is now East Nebraska were particularly beholden to shooting stars and created a sophisticated detailing of the cosmos. They did not merely hold beliefs around meteors, they based their lives and actions on astrotheology (how the cosmos might influence, interact and intersect with life on Earth).

Stars were not only deemed deities in Skidi Pawnee traditions, they were their ancestors, giving new meaning to that now-hackneyed phrase that 'we are stardust': to the Pawnee, shooting stars were more than just molecular structures similar to our own, they were relatives, and shooting stars could be messengers from past or future generations. Ultimately, shooting stars were the children of Tirawahat, the big god, who resided at the zenith of the Universe. Therefore, they were sacred.

In their map of the cosmos, the Morning Star (identified as Mars, but could also be Venus, or occasionally Jupiter interchangeably) was considered to be the first father; and Big Black Meteoric Star (astronomers have since suggested this might be Vega – the brightest star in the north-east sky and perhaps the area from which a large meteorite had once fallen, giving the star its Pawnee name), along with two other deities, held up the sky. The Big Black Meteoric Star controlled the prosperity of autumn harvests, black corn and bears. The star's physical form was imagined as a buffalo* that carried

* Bison are referred to as 'buffalo' in Indigenous tradition.

heaven on its back, and was in charge of ensuring that day passed into night.

The most renowned star lore of the Pawnee is the ritual of the Morning Star, a creation drama symbolising the union of the Morning and Evening Star to ensure the fertility of the earth and the abundance of buffalo. In this ritual, an enemy tribe was raided and a girl from that tribe was brought back to the camp. She was held there for four days, and treated well – like a bull before it enters the ring. No man except her captor could touch her, and if they did, they became the sacrifice. For that reason it was unwise to even use the same spoon she ate with.

A combination of tribal narratives and eye-witness accounts from the mid-19th century vary in the exact motions of the ritual – it might incorporate a sacred 'meteorite bundle', which may contain retrieved meteorite fragments or a Big Black Meteoric stick. But one account from James R Murie, a Pawnee and historian, attests that after days of various rituals and songs, on the fourth day, when the Morning Star rose, the girl would be adorned with a tanned hide as four priests representing the four cardinal star deities sang:

*You now wear the robe belonging to Grandfather Meteorite.*
*You have the covering belonging to Father Meteorite.*

*You have the covering belonging to Uncle Meteorite.*
*You have the covering belonging to Brother Meteorite.*

The girl's body was then painted red on the right side, symbolising day, the time of the Morning Star; and black on the left, to represent night, the time of the Evening Star. The girl then had to ascend a pole made of willow (the Big Black Meteoric Star's associated tree) of her own free will, and the priests would prepare to perform a series of ritual acts.

First a man would come with a burning stick, representing the Sun, grazing the girl with it lightly on her arms and loins. The next man would shoot her through the side with an arrow, striking her heart. The next man representing the Big Black Meteoric Star would make a small incision in the skin above her heart, adorning his face with her blood in four motions. Finally, a man would come with a war club, which he would feint at her four times, over four points of her body, before touching her heart and sending her spirit to the stars, where she would watch over the people for whom she was about to be sacrificed. The girl would then be shot in the back with as many arrows as would fit by every male in the village – including young boys and even babies, helped by their mothers. Her soul would return to Tirawahat, her blood would feed the earth, bless the tribe's dried meat, and these fierce stars that were their gods would be appeased.

# Free Stars

## Norway

In Norwegian myth, we owe Ymir the very air we breathe. Ymir was once a giant, made from the bile, spite and venom of the cosmos. Hell-bent on creating an enemy of the gods, preventing creation, Ymir had to die.

The celestial cow that had once given sustenance to the primeval giant now licked the first god from a salty, rime-covered stone: the cow brought forth the father god Búri. From the loins of Búri came the giant's doom: Odin, and his brothers, Vili and Vé. In the dead of the then starless night, the three brothers cornered the giant in his smouldering home of cinders – the source of all heat in the cosmos. They dismembered Ymir, and as they did, they began to shape Heaven and Earth.

The giant's skin and muscle were fashioned into a ball, creating this spinning planet. From his veins poured blue blood, like that of a king crab, which created our oceans. His bones were chipped into hills and mountains. His hair created our breathing trees. His eyebrows were reserved to create the platform of Midgard, a Middle Earth for humankind, on which his now-earthly flesh could rest. Cracking open the giant's skull, the brothers strew his brains across the sky, creating our 'baleful clouds'.

But what was there to look up to in the night sky? A vacuum of darkness. So, out of his white skull, the brothers created the heavens. Appointing four dwarves to hold the skull high above the Earth they created our cardinal points of North, East, South and West. From the giant's flaming home, they stole embers and cast them into the dark. Some embers were assigned set movements, and others were allowed to roam the sky, free. In Norse myth, these free stars are our shooting stars.

# Akkadian Omens

## Ancient Mesopotamia

A hand presses the fibrous stalk of a giant reed, pulled from the bank of the River Euphrates, into quick-setting clay dug from that same bank. The cuneiform characters the reed forms as it is pressed into the clay tablet spell out an omen that was first written in the sky. The people of the ancient Akkadian Empire, which spanned Mesopotamia towards the end of the 3rd millennium BCE – have not forgotten the guidance of their gods, and this hand hopes to preserve those messages for the future generations who will live under this sky. It writes:

> *If a meteor comes from above the Wagon Star [the Big Dipper] and is bloodstained and passes at the right of the man: that man will see injury.*

When this tablet is broken, hundreds of years later under the rule of King Nebuchadnezzar, this omen and many others will be reproduced as closely as possible. The result was the Enuma Anu Enlil and Namburbi texts – containing some 7,000 incantations and omens. The court astronomer of Nebuchadnezzar I reined (1124–1103 BCE), writes:

> *If a fireball flashes and its flashing is as bright as daylight, and it has a tail like the scorpion the animal while it is flashing, it is a favourable omen, not for the master of the house, but for the whole land …*

A great number of these 7,000 omens concern the observable universe and the fire that falls from it. Over approximately 1,000 years, the cuneiform characters for meteor evolve from left to right.

The Akkadian peoples were knowledgeable about the extraterrestrial origin of meteorites and knew that these heavenly bodies falling to Earth were made of stone and iron. But this is where logic ended and superstition, belief, magic and coincidence converged. Distinguishable from the Fates of Ancient Greece, the trajectory of these stars did not coincide with an immutable destiny for the observer, rather they acted as a warning, or a blessing, but nothing was set in stone. The observer had agency and, if they heeded these warnings, could perhaps avoid or enshrine their fate, as they wished.

If a fallen star (meteorite) fell on a man's house in the Akkadian Empire, he was to expect unfounded accusations. In order to prevent such a situation, he must take that meteorite in his hands, recite an incantation three times, seal the meteorite in clay and throw it into the river, saying, '*Šamaš* [the Sun god], lord of heaven, turn this sign away from me.' Praying to the stars that his sin, like the meteoritic iron, may never return to its home.

Many prophesied outcomes relied on the meteor, comet or bolide's direction of travel, its brightness and whether it was seen during the day or at night, and many of these outcomes concern battles, the fates of armies and lands, and the fall or rise of kings, but they also concerned more domestic matters.

> *If two [shooting] stars in Scorpio flash and move to the western end: it will rain twice in that month.*
>
> *If a meteor comes from above the Wagon Star and that meteor is [as if] dyed red and passes at the left of the man: he will be happy.*

Shooting stars might turn into leather, peas, locusts, bitter barley, clay and dust flies, among many other things. Some omens read more like shipping forecasts:

> *If somebody starts out on an undertaking, and a [shooting] star flashes from the right side of the man to his left side: favourable.*
>
> *If it flashes from left to right: unfavourable.*
>
> *If a [shooting] star behind the man flashes from right to left: unfavourable.*
>
> *If a [shooting] star flashes in front of the man: favourable.*

When I was standing on that island in the middle of the sea, during a pandemic, I can't remember which direction the stars were falling. But I was observing the very same practice as the people of Assyria and Babylon some 3,000 years earlier: venerating the stars and the gods of the night, hoping that by doing so I might encourage a favourable outcome.

# She Who Works Her Will

## Ancient Greece

Favoured as easy sport, quails are flushed out of golden *Molinia* grass by dogs. They soar up and, for a moment, it looks as if they were born to fly, as if they will glide up into the clouds like skylarks. But their rounded wings will only take their heavy bodies *just* far enough from harm, and often, not far enough. After a short distance a covey of quails will tire and fall.

Although Plutarch (46–c.122 CE) credited the Ancient Greek philosopher Anaxagoras (c.500–428 BCE) with the radical, rational theory that meteors were rocks falling from the sky, the Ancient Greek polytheistic notion that the heavens were ruled by the gods was a foundational belief. This is represented in the Ancient Greek word *diopetḗs*. The word was often employed in relation to meteors and meteorites and means 'fallen from heaven' or 'thrown by the gods'. Literally translated it means 'fallen from Zeus', and this was the trajectory of the deity Asteria. To evade the king of the gods' predatory attentions she transformed herself into a ground-dwelling bird so that she would fall from the sky, like a shooting star.

Zeus hunted as he pleased – notoriously capricious and unrelenting with his quarry. He considered women, even those who were Titans, fair game. Residing in the clouds of Mount Olympus, Asteria was the daughter of the Titans Phoebe and Coeus. Her name means, 'of the stars' or 'starry one', and from birth Asteria was associated with the night, with dreams, prophecy and the navigation of darkness. She was the younger sister of Leto, who Zeus had already noticed and impregnated. Following the Titanomachy – the great, decade-long war between the Olympians and the Titans, in which the Titans had been defeated and forced to live in Zeus's kingdom of Olympia – the king of the god's eye fell upon the 'starry one'.

Zeus transformed into an eagle to hasten his pursuit of Asteria. To evade her predator, Asteria also turned herself into a bird, one that is only capable of short, powerful bursts of flight designed to evade danger: a quail. Lone as a planet in the twilight sky, Asteria undertook her first and last flight, her short brown wings carrying her just far enough to parry Zeus's extended talons, before plunging to Earth. Asteria crashed into the blue sea of the Aegean, where she formed the island associated with Delos* from the wreckage of her body.

Owing to her swift descent from Heaven, Asteria would become the goddess of shooting stars, and to this day she is associated with the Perseid meteor shower, named after her husband's constellation, Perses. Before Zeus's attack, Asteria and Perses had shared a child, Hecate. Like her mother, Hecate became a goddess of night, beloved of the witches of Thessaly and kindred to crossroads, herbal drugs – poisons and balms, magic, the Moon and the dogs that howl at it.

Hecate's name means 'willing'; rather than 'obliging', it translates as 'she who works her will', like her mother. As the poet Callimachus wrote of Asteria, 'since like a star thou didst leap from heaven into the deep moat', she is a reminder that a fall from grace isn't always a bad thing, there can be power in it.

* In some older traditions of this myth, Delos has another name, now attributed to another Greek island, Ortygia, meaning 'quail island'.

# Zmej

## Eastern Europe

Woods and forests have been described by acclaimed naturalists as the 'subconscious of the land', a place where imagination thrives and where the line between dreams and reality becomes liminal, vertiginous. Is it any wonder then, that with 43 per cent of its land covered by woods and forests, Belarus has a rich culture of folklore, myth and superstition, images and meanings created for the minutiae of everyday.

Here, the direction of a meteor's travel, their trajectory, line of flight and style of flight, were all open to interpretation. If a shooting star fell straight, the deceased

(whoever they might be) was a righteous man; if it fell at Yule, it might be an unbaptised baby freed from Hell to have a good time; if the falling star flared, there would be a good harvest.

Then there were the fireballs and bolides, the meteors that exploded in Earth's atmosphere and transformed into a Zmej or a Tsmok, Slavic dragons – gold chasers, fire starters and rakes. The Zmej was a fiery meteorite, and if one was seen twisting and turning through the sky, it was believed that folk should put milk out in their gardens and porches to ward off the fire from their houses. A fire begot by a Zmej could only be quenched by milk, water would only encourage its burning. If you laughed at such a star, it would seek vengeance and burn down your house. There were other reasons that one should beware a Zmej flying over your home: it might be carrying gold for a witch or wizard, and if the starry dragon flew over your house, your identity as an accursed person would be revealed.

No house was safe from the Tsmok, another serpent-like creature associated with fire and meteors. The Tsmok flew down chimneys in the dead of night and, as it landed in the smouldering grate, it would appear as a handsome young man. The young women of the house might come down to see what the commotion was and, if they did so, they were destined to fall in love with him. Once enamoured, the Tsmok would come to their rooms at night and drink milk from their breasts, delighting vampirically as he drained their strength day by day, until – in most cases – they died.

If she survived, the afflicted girl might be regarded as a witch, which rarely turned out well for her. If a woman in the village gave birth to an unhealthy child, she may be accused of sleeping with the Tsmok. To be the beloved of the Tsmok was to stand alone in an isolated community, with only the trees for company.

# Hanging Stars

## England

'Do they burn like the jewels of my manuscript?' The monk wonders as he looks beyond the mist of the mesosphere and into the layer of fixed stars, held up by two dutiful angels. Only occasionally do stars slip through their fingers and fall to Earth, perhaps during a momentary lapse in concentration, or when an angel needs to rest a weary arm.

From what little the monk remembers of the disregarded oral folklore of his Pagan ancestors, falling stars demonstrate the boundary between worlds and the descent of spirits. But he knows better than that now. He knows that some of these falling stars – like the Perseid burst he is awaiting this August – are mournful little sparks, the burning Roman coals that had cooked the flesh of the martyred St Lawrence, or were they the saint's tears? Different texts said different things, but hadn't that always been the way of things.

Seen through the eyes of the monk, the Moon hangs luminous in her white shroud above the monastery. She has been pulled to her evening station by a carriage of white oxen, and the Sun to his rest on his chariot of stallions. The Earth has been turned on its axis by the two angels, cranking up our orb's windlass for the night. The monk stands on our planet in its nocturnal position, in the midst of the Dark Ages, in the middle of the night. And night is truly night here in deep history, in deepest Wessex.

At this time during the Early Middle Ages between the 5th and 10th centuries, Britain's history, religion and superstition were in the process of being welded together within the silver and gold of the illuminated manuscripts painstakingly produced at monasteries. This monk lived under the reign of King Alfred the Great, sometime during the late 800s, and was one of many that were tasked with compiling a detailed, year-by-year history of Britain following the departure of the Romans, called *The Anglo-Saxon Chronicle*. The monks drew from various documents and other source material called the 'common stock', written by earlier, often unnamed hands.

By candlelight, the monk has just copied out the observation of a comet that appeared 200 years before: the first comet recorded in the skies above England. On vellum, he duplicated old records reporting that a comet appeared in the sky during the month of August in the year 678 CE, and that the comet had 'shone for three months each morning like a sunbeam'.

*First cometary reference in The Anglo Saxon Chronicle 678* CE

**Her ateowede cometa and scan III mon as aelce morgen swilce sunne beam.**

'HER ATEOWEDE COMETA AND SCAN III MON AS AELCE MORGEN SWILCE SUNNE BEAM.'

**'In this year appeared the star called comet, in August, and shone for three months each morning like a sunbeam.'**

Stars like this 'cometa' – those that appeared suddenly and hung in the sky for days, weeks or months – were the stars to worry about. Comets were an augury of ruin: a sample of the Hell yet to come from Heaven. They were seen as part of the arsenal of a god who loved his creation conditionally, a god who relished his interventions and set them aflame for a trembling audience of monks to depict. Comets in Britain were 'like flaming swords' the 12th-century historian William of Malmesbury tells us, and could be seen as heavenly creatures, divine dragons migrating across the illustrious, astral planes. This was a time when the skies above were alive; they were a playground for monsters and seraphim.

Comets threatened all life, but particularly the lives of those seemingly ordained by God: bishops, kings and princes. As these personages rose high, they were especially vulnerable to a comet's scorching train. That downfalls arrived on the wings of shooting stars was a belief that persisted throughout Britain for thousands of years. In 1907, a meteor exploded above Nottinghamshire, and a local was heard to say, 'Ah! We shall hear of some big man's death.'

As for our monk in the late 800s, he looks out at the night sky over Wessex, he prays that he will not see the 'cometa' now on his mind, that he will not see the fall of his king, or another war. He gazes at the stars he so diligently depicts in the ornamentation of his manuscripts during the daylight hours, the work that has blighted his vision and hampered his ability to view the distant heavens; his myopia is the reason starlight arrives like four flashes of illumination in his eyes, like a cross.

# Everywhen

## Australia

The origins of the mushroom are in the sparkling firmament, and they should never be eaten: so follows the lore of the Arrernte people of Australia's red Northern Territory, where a mushroom is a fallen star. This forms part of the Arrernte's unique worldview they call Altyerre, which literally translates as 'everywhen', but is also known as Dreamtime or The Dreaming, a tradition the Arrernte share with all Indigenous Australians.

This Dreamtime worldview encompasses a spiritual perspective of creation stories, of past, present and future, and of the interconnectedness of all things. And it was the Altyerre oral tradition of the Arrernte that led to the rediscovery of an impact crater, now colloquially known as the Dreamtime meteor crater.

The Arrernte's origin story of the formation of Tnorala, or the Dreamtime crater, is that women were dancing among the glitterballs of the Milky Way. Unable to find suitable childcare for the evening, one of the women brought her baby with her. Her arms soon tired of carrying the baby; her husband was nowhere to be seen. Longing to stretch and trace her fingertips through the residual dust of supernovas, the woman placed her child in a basket and continued to dance. The baby fell out of the basket, tumbling through space, descending into the exosphere, creating bright fire through the thermosphere, before hitting the Earth, forming the crater. When the basket fell on top of the baby, the force of that impact pushed the surrounding rock upwards in a perfect circle.

In reality, the crater is thought to be the result of a comet impact over 142 million years ago, but its extraterrestrial origins were deduced by the Arrernte and perpetuated through oral lore long before its official 'discovery' by the British-born explorer Ernest Giles in 1872.

To many of Australia's Indigenous tribes, falling stars have generally been associated with bad magic and malevolent spirits that might be looking to steal babies, which they loved to eat, or else these stars fall to Earth in order to tickle people to death. But, to the Karajarri people of coastal Western Australia, a 500-million-year-old giant nautiloid shell supports the dome of the night sky. Within this prehistoric sailor are a million tiny shining nautiluses. To the Karajarri, a falling star represented one of these marine molluscs dying and falling from its shell.

In Arnhem Land, in the north-east corner of the Northern Territory, the Yolŋu people see a shooting star streaking across the night sky as a lit beacon. It is a 'canoe of light', signalling from the heavens that the person they were mourning has crossed the great divide and arrived safely on the other side. To the Boorong of north-west Victoria, the Lyrid meteor shower is caused by a celestial, quail-like bird, known as a Neilloan, scratching at the firmament until it sparks as the bird builds its nest.

And there is another Arrernte legend, that water shouldn't be taken from the pools near craters in the Northern Territory because of the fire devil. The fire devil travelled from the Sun and slammed into Earth, forming the huge craters in the area, and if you drink from them, he will fill you with iron.

The strange formation of one of these craters was noted by a cattle rancher in 1899; in a letter to an anthropologist he wrote, 'To look at it I cannot but think it has been done by human agency, but when or why, goodness knows.' Though the Arrernte knew that these craters had their origins in space, the nature of the site remained a mystery to those outside the Indigenous community until 1921.

That year, a local prospector, J M Mitchell, was taken to the craters by an Arrernte guide, who was too frightened to visit the site itself and would not camp anywhere nearby. The craters were soon named the Henbury Craters after a nearby cattle station, and a few years later, another prospector realised that the craters contained iron-nickel fragments, debris from the impact of an octahedrite, an iron meteorite. It is likely that these fragments were the origin of the belief about being filled with iron if someone dared to drink from these craters.

As with many superstitions and folk beliefs, the Arrernte legend connected with the Henbury Craters is rooted in an event that was either witnessed first-hand, or deduced through similar impact events and structures, thus forming a collective memory of destruction passed down through generations, covering perhaps as many as 4,700 years. Primarily, these legends were created to protect the community: they served as a caution.

Evidence of three meteor impact events have been found in Australia using a combination of Altyerre tradition and Google Earth. These discoveries are examples of a relatively new field of science known as geomythology, using ancient lore to make real world findings – from impact craters and fossils to earthquakes and volcanoes – previously unknown to Western minds. Some findings are as old, or older, than the giant nautilus, which now only exists in the Karajarri heavens.

# Astral Migration

## Southeast Africa

Since the dawn of time, the feet of celestial cattle have broken through the floor of the sky, their hooves sparking against the metal of heaven, tearing streaks of light that soon fill with the mud of night.

In the culture of the Indigenous Zulu peoples of Southeast Africa, shooting stars do not have the kismetic connotations traditional in many cultures when assigning meaning to natural events. In Zulu lore meteorites have customarily been seen as neither good nor bad, they are merely an indication of the heavens getting on with the same business as the living: the sky above is a great blue rock that experiences cycles and seasons in the same way as life on Earth.

The Zulu describe the Milky Way as the main entrance to the cosmic cattle's enclosure, ruled by a great celestial chief, an *iNkosi*, and the stars are where his animals have trampled through the rock after a heavy rain in search of new grazing. Falling stars occur when the cattle are in a hurry to reach this grazing ground, and drag their feet through the floor of the sky. 'The light is seen before the mud fills it again,' an elderly Zulu woman told Brian Fraser, a visiting astronomer, in 1995. 'So it is clear *iNkosi* there has cattle.'

To the Zulu, cattle are sacred, and their stables form the centre of Zulu social life within the *umuzi*, the family settlement. As they also serve as a connection to ancestral spirits, cattle are not merely a lifeline of material sustenance, but symbolic of spiritual health. In Zulu tradition, doors and openings face towards where the cattle graze in the evening. Cattle have bound families through marriage, been exchanged as *lobola*, a dowry, and legitimised children. They are regarded as intermediaries between the material and spiritual plane, and so are sacrificed to bring the dead closer to the living.

If these horned intermediaries close the gap between 'here' and 'there', a shooting star, a tear caused by a celestial hoof, might be seen as a reassurance that a departed soul will not be without the sound of their cattle, continuing the rhythms of life on another plane.

# The Flight of Souls

## Mesoamerica

The Earth was a turtle. A 'ghost-patterned' snail held a region in the Underworld. The Milky Way was a sky rope, and the meteors that fell from it were studiously observed by the Mayans, the ancient civilisation in Mesoamerica greatly concerned with the study of time and stars. The Mayans had a comprehensive, sophisticated understanding of constellations; of lunar and solar calendars; and the cycles of meteor showers and comets.

Revered among the Mayan civilisation were the 'seers', who were believed to have the capacity to master dreams, and transform into animals, lightning, and even into

comets and bolides. Their perceived ability to perform these transmutations was evidence of their powers.

To become a seer, a novice priest must first be initiated. These ceremonies typically took place in a cave, where a dream-state was achieved, either naturally or through the use of hallucinogens extracted from the root of a water lily. The resulting dreams would typically concern comets and meteor showers and their calendric appearance; the Moon and its lunar cycle, which governed both the living and the dead; and a meeting with a room of their god-like ancestors, who were subject to something akin to the concept of eternal recurrence.

Such was the importance of comets in Mayan cultures that they anticipated their appearance in the night sky, were careful of the lunar phases to ensure optimum observation, and noted the various kinds of shooting stars, and their different behaviours, taxonomising them with a variety of names:

**Ik-om**: comet
**Hal-al ek**: running star
**Ek ku lubul**: bolide
**Chaman tz 'utan**: a magician's cigar (small comet)

That souls would be reincarnated was part of the Mayan worldview, and the time it took for a soul's reappearance in the material world was governed much like a pregnancy, in accordance with the lunar calendar, but it was also governed by the appearance of meteors or comets.

If a Mayan woman was pregnant, it was important that she find that baby a soul. To do this, the woman had to meet her mystical father who would appear to her in the night sky as a comet or a meteor. She would climb the steps of a ceremonial pyramid, no later than one month after discovering she was pregnant, and look up at the sky night after night until she saw this soul-bearer burning through the black, and only then would the celestial father bestow a soul upon her unborn baby.

# Water Chestnut

## China

A water chestnut fresh from the earth is dark brown, round and can be held between three fingers – approximately the size of a medium conker or horse chestnut. But the water chestnut is not a chestnut. It is the tuber-like root of an aquatic, sedge-like grass associated in Chinese lore with stability, good fortune and, on one fateful day, with falling stars. They have been used in medicine for boosting chi and alleviating numbness, and to add crunch to recipes since the Han Dynasty (202 BCE–220 CE).

The Han Dynasty is widely considered China's imperial Golden Age, as leaps were made in art, culture, social conditions, inventions (such as paper) and astronomy. It was a time when illustrations of comets were made in astrological manuscripts, such as *The Divination by Astrological and Meteorological Phenomena*, which was compiled by astronomers during the Western Han dynasty (202 BCE–9 CE). It is one of a number of manuscripts discovered in a tomb in Hunan in 1973 known as 'the silk books', as the illustrations and diagrams were drawn on silk. These form some of the oldest, most comprehensive, and most accurate observations of comets and meteorites in human history.

Early Chinese records of comets and meteors from this time include a variety of descriptive names: from 'bushy star' and 'sweeping star' to 'heavenly intermingler', 'sailing star' and 'candle flame star'. They may also have been called 'Tiangou'.

In Chinese mythology, Tiangou was a celestial black dog that resembled a meteor and was held responsible for eclipses. It has been posited that the Tiangou's association with meteors arose from a recorded meteor strike in the 6th century CE, the black dust from which is believed to have obscured much of the light of the Sun. On one ancient manuscript the heavenly black dog is depicted prowling around the Sun, on a cloud of his own making.

The notion that a meteor strike might create clouds of dust around the Sun – which could lead to famine and plague – led comets and shooting stars to be regarded as a sign of imbalance in the cosmos. Under the theory of Wu Xing, the Five Elements, comets were thought to signify an imbalance of galactic yin and yang and were largely considered a portent of disaster.

This belief is supposed to have been realised in one of the heaviest meteor showers recorded in Chinese history. In 1490, a meteor made a noise like 'a flock of cranes in flight' and then spewed meteorites down onto the population. Though it has been disputed, one surviving source from the time claimed the shower resulted in hundreds of deaths. The larger falling stars were compared to 'goose's eggs', and the smallest to 'water-chestnuts'.

PART FOUR

# When the Star Falls on the Moon:

## *Shooting Stars' Influence on History*

# The Welkin Aflame

It was a wet November morning the day I finally made it to the Holy Island of Lindisfarne. The mist from the North Sea became indistinguishable from the low-lying cloud above the light sand beach. Cormorants questioned the silver water of the rising tide and the wind blew damp gusts against a great red-brick arch, still standing despite much of the island's monastery now lying in ruins. Like so many others over the centuries, I had crossed the causeway to this tidal island to visit the monastery where the hermit St Cuthbert had once lived. This was the same reason, around 1,300 years earlier, that the Venerable Bede, one of England's greatest chroniclers, had also made this journey, the same year he finished his prose version of *The Life of St Cuthbert*.

A monk and scholar, Bede chronicled the rise and fall of kings, the lives and deaths of saints and holy men, the wars and disputes, the friendships and allegiances of the British past, and thereby installed a squint into those bygone years. Over his 61 or 62 years, Bede wrote around 40 books on different aspects of knowledge, from history and nature, to poetry and astronomy. 'They are born suddenly, portending a change of royal power or plague or wars or winds or heat,' the Venerable Bede wrote of comets in his natural treatise *On The Nature of Things*.

For those in medieval England, including Bede, interpreting the omens of comets became less superstitious practice and more a logical case of taking what one could gather from coincidence. Over many hundreds of years, comets and shooting stars undeniably seemed to coincide with England's misfortune, and that of its rulers.

The ancient, hegemonic kingdom of Northumbria, where Bede served out his adult life, can hardly be said to have been a stable kingdom during his lifetime. Bede was five years old in 678 CE, when the first mention of a comet appears in our written record and he would later frame this as the comet that heralded the devastating defeat of the Northumbrian King Ecgfrith by the Picts at the Battle of Dun Nechtain in 685 CE. What Bede does not mention is that there was an appearance of Halley's comet – visible across all of Europe – in October 684 CE, just seven months before King Ecgfrith's undoing.

By the time Bede wrote his most famous text, *Ecclesiastical History of the English People*, in around 731 CE, he had seen the hand of God in the sky several times, and learned from an early age that its fire surely presaged upheaval. The year 729, when two comets appeared, would prove to be a pivotal one. 'In the year of our Lord 729, two comets appeared around the Sun, striking terror into all who saw them,' Bede wrote. 'One comet rose early and preceded the Sun, while the other followed the setting Sun at evening, seeming to portend awful calamity to east and west alike. Or else, since one comet was the precursor of day and the other of night, they indicated that mankind was menaced by evils at both times.'

On cue, St Ecgberht of Ripon, a monk, nobleman and celebrated pilgrim, died immediately after Easter Mass. The man was 90, but still, his death was believed to have been foretold by the appearance of the comets, particularly because the Northumbrian King Osric also departed this world a few days later. Bede writes that these comets were 'pointing their fiery torches northward as though to set the welkin [heavens] aflame'.

Long after Bede had ceased to document them, comets would continue to thunder through the cosmos and seemingly rain their spiteful fire down on England, and the kingdom of Wessex, especially. In 905, the popular Prince Æthelwold's attempt to seize the crown of Wessex was doomed by the appearance of a comet.* In 975, Edgar of Wessex's long and peaceful rule came to an end after the appearance of a comet, that is now known to have been Halley's Comet, and a famine ensued. It was recorded in *The Anglo-Saxon Chronicle*: 'Then too was seen, high in the heavens, the star on his station, that far and wide wise men call – lovers of truth and heav'nly lore – "cometa" by name. Widely was spread God's vengeance then throughout the land, and famine scour'd the hills.' Then, there was the reappearance of Halley's Comet from on high in 1066, its perceived divine power foretelling England's king, Harold of Wessex's, fall and William of Normandy's God-intended victory.

Though it was the ascribed meaning of comets, and the behaviour influenced by this meaning, that affected the outcome of events, or the way events were perceived and correlated, the coincidence surrounding comets and

* Some versions record his defeat as 902 CE and the timings may well have been manipulated to align with the comet.

fateful happenings in the British Dark and Middle Ages was written with fire in the sky for all – born high or low – to see, and that was a power they held the world over. To rulers and politicians around the world, a shooting star would continue to be deemed a godsend, a caution, a seal of approval, a divine ruler or damnation from on high, and those in positions of power learned to ensure these portents served their cause.

As they are visible to all, regardless of status, shooting stars were the spin-doctors' first instance of a message in lights. Written on high, bigger and brighter than any illuminated billboard or television screen, political and religious advertising campaigns throughout history have used shooting stars to propagate social values, mobilise troops, garner or divide loyalty, and spread social and religious messages. The degree to which these portents changed life for the ordinary person might have been relatively minimal at the time, but as time gathered ground over the centuries, ruined monasteries and put all god-fearing men in their graves, these events have had a cumulative effect on how we all live. In such uneasy times as ours, who's to say what our rulers and leaders might think or fear in their secret hearts if they looked up and, like Bede, saw the appearance of twin comets setting the welkin aflame both day and night.

# A Smoke Banner

In 6th-century India, it was important to know the order of things. From Heaven to Earth, scholars were neatly taxonomising the world around them and the firmament above. White meteors were the concern of gods and kings; red meteors should largely preoccupy the warriors; yellow meteors were to be noted by the merchants; and black meteors were the office of the labourers. So followed the sequence of the cosmos according to the Indian astronomer Varāhamihira. Like many citizens of the early medieval era, Varāhamihira also believed that the appearance of a comet or meteor could offer clues as to the direction of future events – in particular, if a meteor shower (rather than a lone meteorite) rained from the sky, it was the harbinger of destruction.

About a thousand years after Varāhamihira recorded his taxonomy of falling stars, in the region of what was then known as Satisara, now the Kashmir Valley, a medieval continuation of the 12th-century Sanskrit chronicle known as the *Rājataranginī* (*'River of Kings'*) tells us that in 1532 the people of Kashmir had recently been under attack from a Kāskār army, warriors from Afghanistan and the Khyber of Pakistan. Shortly afterwards, a comet – known in the ancient Vedic scriptures of Hinduism and Sanskrit as *dhumaketu* (a smoke banner) – appeared in the sky: '... a comet was seen continually in the sky ... Thus the ravaging of the Kāskārians took place. Stars fell from the sky on the fields where the full harvest of rice was ripening, and a comet became again visible ... this calamity devoured the grains; and there happened a great famine, the destroyer of food.'

The subsequent famine was so severe that the people of Kashmir resorted to eating those who had already succumbed to hunger. A meteor storm may have damaged some of these rice crops, but this famine of 1533 was largely caused by unseasonable summer rain, snow and flash floods. As Shukar's *Rājataranginī* demonstrates, at the time the decimation of the population was viewed as the sorry vindication of Varāhamihira's belief that meteor showers were the harbinger of destruction, because that was the order of things.

# Planta Genista

*There appeared a star of wonderful Magnitude and Brightness, darting forth a ray, at the End whereof was a Globe of Fire in Form of a Dragon, out of whose Mouth issued forth two Rays; one of which seemed to stretch out itself beyond the Extent of Gaul, the other towards the Irish Sea, and ended in seven lesser Rays.*

This comet is described by the medieval bishop and chronicler, Geoffrey of Monmouth, who claimed to have compiled the legends of King Arthur from an ancient tome in the British tongue, given to him by the Archdeacon, Walter of Oxford. Monmouth published this work under the title *History of the Kings of Braitain* in around 1136 and the dragon-headed comet concerns the prophesy of the sorcerer, Merlin. Upon being asked the 'signification of the star', Merlin foretells the death of a king, and the rise of his brother, Uther Pendragon ('Pendragon' meaning 'head dragon' in Welsh). The comet heralds the invasion of the Saxons and the birth of Uther's son, King Arthur.

Arthurian legend is exactly that: myth and legend, with some rocky footing in reality. The obscurity of Arthur's existence has made him a malleable figure and over the centuries his lore has been used to shape the British identity, through which a sense of patriotism could be inspired and then employed to rally troops, pacify the disgruntled and legitimise a dynasty. The legendary figure that this comet heralded shaped public opinion and, ultimately, British history.

The story of the House of Plantagenet began in Anjou, France, with Geoffrey V and his queen Matilda, the daughter of Henry I, the fourth son of William the Conqueror – whose conquest of England was also anticipated by a comet. Geoffrey wore a sprig of broom, the Latin name for which is *Planta Genista*, the origin of 'Plantagenet'.

Geoffrey's young son Henry 'Curtmantle' lived in England during the reign of Henry I, under the protection of his uncle Robert, the Earl of Gloucester. Henry lived in Bristol for a time, where he was likely regaled with the tales of King Arthur. In part, due to Bristol's proximity to Wales, where the legend of King Arthur was held particularly dear; and also because Geoffrey of Monmouth had dedicated *The History of the Kings of Britain* to Robert.

Political and military machinations led to Henry being crowned King of England in 1154, aged 21, and the formative years of that young Norman in the West Country, and the legends he heard there, would prove to be the making of a British dynasty. His awareness of his new realm's intense connection to Arthurian legend inspired Henry and his wife, Eleanor of Aquitaine, to link their ancestry to King Arthur's. This was achieved through the suggested inheritance of a common nature rather than a direct blood lineage.

The Plantagenets thereby established themselves as King Arthur's spiritual heirs, and they ensured their courts reflected the chivalric values of the Arthurian court, creating knightly orders, a sense of courtly romance and appropriating pre-existing symbols of the realm such as lions and dragons to further their association with the myth. The pair also patronised writers to ensure translations were made of the legend from Monmouth's Latin into Old French – the language of the court and aristocracy in this Anglo-Norman land. This secured the widest possible audience by continuing the myth's popularity in Europe (it was the Norman, Robert Wace's, translation of the myths that introduced the notion of a round table to symbolise equality among the Arthurian knights).

The Plantagenets had hoped these efforts might placate the indigenous Brittonic belief of the Welsh. They were loyally awaiting King Arthur's mystical return – known as the Breton Hope – and continued to resist Henry's attempts to assert his authority. The supposed discovery of Arthur's body shortly after the death of Henry II was intended to encourage the acceptance that Arthur, the 'one true king', was indeed dead, and to instead accept the Plantagenets as the next best thing.

Through coincidence or providence, Merlin's prophecy at the sight of the comet was largely fulfilled. The self-proclaimed descendants of Pendragon did, for a time, end up ruling from Gaul across the Irish Sea. The Plantagenet dynasty embedded itself into the country once known as Albion and used Arthurian myth as a model of behaviour for themselves and their subjects. They continued to rule England and parts of Wales, Scotland and Ireland for 331 years, until the death of Richard III in marshland in Leicestershire at the Battle of Bosworth Field in 1485.

By then, these legends had become entrenched in the British psyche and the behaviours, attitudes and qualities they propagated have persisted as part of the British identity. 'Make haste, therefore, most Noble Uther,' Merlin tells Uther Pendragon. 'The victory will be yours, and you shall be king of all Britain. For that star, and the fiery Dragon under it, signifies yourself.'

# House of Fire

It is 15 March and a cigarette, a few coins and some bunches of echinacea, pink roses and gypsophila have been carefully laid on the mound where a funeral pyre once burned. Now, at dusk, as the first evening stars appear and the last of the swifts scree, it is quiet among the ruins on the Ides of March; the sounds of scooters and voices from neighbouring streets rub shoulders with the funerary atmosphere. Earlier, a few hundred Romans and tourists were here to pay their respects to the murdered emperor, Julius Caesar. We are standing at the Temple of Deified Julius or the Temple of the Comet Star.

In 44 BCE, a comet appeared during the Ludi Victoriae Caesaris – games held in honour of Caesar's military triumphs and to commemorate his life. Four months earlier, on 15 March, Caesar had been assassinated by a group of conspirators from the Roman Senate who feared his dictatorial ambitions. Now, a comet shone in the sky, where it remained for seven consecutive days. Much like the Tsuchinshan-ATLAS comet of 2024, the comet of 44 BCE was also visible to the naked eye and appeared just before sunset, allowing the plebeians of Rome to see for themselves – Julius Caesar wasn't just any emperor, he had been a divine ruler. According to the Roman historian Suetonius, as the comet's appearance also coincided with Caesar's birth month, July, it was believed that the comet was a visual sign of Caesar's soul ascending to heaven.

After Caesar's assassination, years of disruption and civil war ensued until Caesar's great-nephew and adopted son, Octavian, as the Emperor Augustus, took control in 27 BCE. Augustus proceeded to take control of the narrative surrounding Caesar's life and after-life, reminding the public of the comet that had appeared at the behest of Caesar's soul.

To encourage the public to his way of thinking, Augustus did what the Romans did best: he built a temple, a house of fire, the Temple of the Comet Star. Ovid writes that above a mural of Caesar in the temple, a fiery comet was attached to his head, 'To make that soul a star that burns forever / Above the Forum and the gates of Rome'.

That same year, Augustus had a coin minted featuring his laureate head on one side, while the reverse featured an eight-pointed star that represented Caesar's comet and Augustus's divine connection to his predecessor. In persuading the general public that the comet signified his adopted father had been divine, chosen and ordained by the gods to be emperor of the Roman populace, so it followed that Caesar's son was also chosen by the gods. With some poetic licence, and the heft of an empirical propaganda machine, the comet's appearance helped smooth the long road of Roman history for the next 400 years at least.

# Who Can Tell Good From Bad?

A farmer stands in the green fields below the Huangshan mountain range in Anhui Province, eastern China, and looks out across the endless paddies, stretching beyond the rocky horizon, towards which his horse has escaped and bolted. The farmer believes that losing his horse in such a way is a manifestation of bad luck, until the horse returns, with more horses, changing the farmer's fortune from bad to good. The farmer's son breaks in one of the new horses and to test its schooling, he takes it for a ride. Still green, the horse bucks causing the son to fall and break his leg and their fortunes change

– the old man feels bad luck sink into his bones once more. A few weeks later, a war is announced and men of fighting age across the land are called to battle, and to perhaps lose their lives. With a broken leg, the farmer's son is spared the draft and their fortunes change anew.

So goes an ancient Taoist parable known as 'Sāi Wēng Shī Mǎ' ('The Old Man Lost His Horse'). Its central tenet is: who is to know good from bad? Symbolising life's interconnectedness and how one perceived good event can lead to a bad one, and vice versa, it urges us not to judge events prematurely, and to let things unfold as they may.

Unlike Julius Caesar and his chosen son, for many in power throughout the ages, a shooting star was not perceived as a ruler's elevation to divine status, but rather an ominous presage from the divine: it was a reckoning, and could herald a king or emperor's judgement day. One might question how much that perception had to do with a ruler's conscience, but regardless, the power of a falling star over the monarchs of our past was immense.

As comets and shooting stars were almost always considered a bad omen, Chinese emperors employed observers specifically to look out for them, so they could be forewarned of imminent disaster. This is partly why ancient Chinese astrological observations comprise the most extensive records in existence.

In one of several examples of celestially induced abdications in China, Emperor Ruizong of Tang, the ninth emperor of the Tang dynasty, transferred power to his son after the appearance of a comet, now known to have been Halley's Comet, in 712 CE. For Emperor Ruizong this comet signified a need for introspection – an imbalance of chi in the cosmos was, perhaps, his fault. He went about purifying his ways. Beginning with a period of fasting, he ordered his ministers to work carefully to analyse any missteps they may have made and look to history for answers as to how best to proceed.

The Chinese philosopher Confucius (551–479 BCE), regarded 'Ren', or 'co-humanity' or 'humaneness', to be the highest virtue a person could possess. Ren encompasses love, empathy and the notion that all human beings are interconnected. Ren also happens to be a central pillar of Confucian political theory: an emperor with the Mandate of Heaven should be altruistic, someone who leads by moral example and prioritises the welfare of their subjects. But Confucius also believed that true virtue is demonstrated by acting with principle, even if this action is unpopular, as someone who looks to achieve popularity through pretence is a 'thief of virtue'. Emperor Ruizong will likely have had these teachings in mind as he was forced into action by the comet.

Abdicating in favour of his son, who became Emperor

Xuanzong, enabled Ruizong to evade misfortune, while retaining governing powers – a detail enshrined in the change of rule at the behest of Ruizong's sister, Princess Taiping. With these artful manoeuvrers it initially appeared as if the emperor had saved himself and his family from a potentially tragic fate.

Princess Taiping, however, was wily, as one needed to be in any imperial court, and had her own designs.

Because Taiping had no respect for her nephew, the new emperor, and felt her sons, and even herself, could do the job better, she rallied her powers and planned a coup against Emperor Xuanzong by poisoning his aphrodisiac. When Xuanzong heard about the plot, he ordered her co-conspirators and family to be killed. Princess Taiping fled to the mountains, perhaps hoping to let things unfold as they may, or that this would give her brother time to intercede on her behalf. Ruizong did intercede and ask for Taiping's life to be pardoned but his son refused, and when she re-emerged three days later, Ruizong felt fortune had forced him into a corner. He handed over all his remaining power to his son, thereby forcing Princess Taiping to take her own life.

After her death, it was a happy turn of events to discover that the princess had so much treasure, property and livestock, that it took officials three years to consolidate her accounts. Who is to tell good from bad?

# Uncommon Star

It was a November night in 1577 and there was fire in the sky where there should have been darkness. A star was gunning it through the vault of black, exhaling a radiant white coma behind its moon-like head and setting the sky alight above London's forest of high spires.

At Richmond Palace, Queen Elizabeth I watched the star's pageantry with the suspicion she reserved for the marriage processions of her ladies-in-waiting. And, like the first days of those marriages, it was too early to tell whether this star's transit through her reign would be smooth, or wont for meddling.

The Queen's advisors were in a panic – comets such as these were portents of doom for princes like her. They prophesied the downfall of kings. 'What does this mean?' they whispered to one another – to her – as they crossed themselves and tightened their ranks. 'What does this mean?'

The Queen shook off her advisors and their anxieties; instead, she would demonstrate the courage they lacked, and that they would later write in praise of – courage that the oldest among them had not seen even in kings. It is not that Elizabeth didn't believe in divine providence and portents, it is simply that she saw things differently: even the hand of God could be open to interpretation, if it were prised open.

She summoned a man who saw things as she did: her astrologer, schemer, naval advisor and alchemist, John Dee. The comet, Dee told the Queen, was 'a portent not of Elizabeth's destruction but of her elevation,' as Benjamin Woolley writes in *The Queen's Conjuror* (2002), 'and the fulfillment of a destiny greater than even she might have imagined.' This comet would become known as 'the uncommon star' and it would change the science's understanding of the Universe.

When the Danish astronomer Tycho Brahe first saw the comet in Prague on 13 November, it was just before sunset, and he was headed home after a long day's fishing. Brahe's subsequent measurements of the comet's parallax led to the acceptance that comets and

meteors were, indeed, a physical reality rather than celestial apparitions, tricks of light. The findings would influence Brahe's assistant, Johannes Kepler, who in 1609 demonstrated that planets had elliptical orbits, thereby challenging scientific belief that had stood still since Aristotle in the 4th century BCE.

At this time, Dee and Elizabeth's astrologer, Thomas Digges, were still operating under the Aristotelian model, but they still encouraged rationality over superstition in order to mitigate the terror induced by the comet, as they had done five years earlier when a brilliant star in the constellation Cassiopeia had caused uproar. Dee had written then, 'Is it not commodious for man to know the very true cause, and occasion naturall? Yea, rather, is it not greatly against the Souverainty of man's nature, to be so overshot and abused, with thinges (at hand) before his eyes?'

Once Elizabeth had consulted Dee, she went to her window to view the comet. Looking at the sky, she decided to gamble. She had her windows flung open, letting in the cool dark air. She then shouted out into the night, '*Lacta est alia!*' ('The dice are thrown!'). This signified a point of no return for Elizabeth. She had decided to be confident that the future she was about to enter was not cursed by God but rather under His protection, a protection that could subdue even the brightest and most penetrating beams of stars.

In his *A defensatiue against the poyson of supposed prophesies*, Henry Howard, Earl of Northampton, wrote of this moment, 'Behold a woman and a Queen, which seem to be the kinds and callings, upon which the Comets (if Astrologers speak truth) are wont to prey: and yet not only she relenteth not to common fear, but insulteth rather upon common folly. Yet her contented mind [...] were enough to verify that proverb which hath been rife of old, *Sapiens dominabitur astris*, "the wise subdue the stars".'

Elizabeth had flipped a long-held perception of the comet: from signifying a cosmic discord that rang down to Earth, and had rung through the bones of the English kings these stars had lain waste to before her, into the manifestation of a divine blessing that would grace her reign for another 23 years. In deciding instead to announce the comet as a visible sign of God's protection over her, her rule and her country, Elizabeth I – adept since childhood at keeping herself safe in dangerous situations – performed one of the best PR stunts in history.

# Black Stone

It stands at the end of a very long road, a road some may have waited their whole lives to walk down. Through the desert, under the beating sun, they will walk, step by step, until they reach it: black and absolute, like the end.

This is the stone that Mohammed set into the wall of the Kaaba, the huge black and gold shrine in Mecca, five years before his first revelation. He sealed it with a kiss, which many pilgrims now emulate, if they can, as they pass its smooth black face. It is set within solid silver into the eastern corner of the Kaaba, and it is this stone that dictates the direction of prayer for Muslims around the world.

The Black Stone – now eight visible fragments held together – marks the site where the sacred world meets the profane. Though it is not known if it is a true meteorite (it could be agate, it could be tektite – a black molten glassy ejecta from a meteorite impact, or it could be meteorite) its believed extraterrestrial origin consolidates its role as a link between Heaven and Earth. It being witnessed as falling from the sky may have been the impetus for the Black Stone's known veneration by the pre-Islamic peoples of the southern Levant, living at a time when interesting or curious rock formations were deemed to have special significance in many cultures. To this day we might collect stones on beaches or from riverbeds to thumb in our pockets, if only for a while. 'It gives us a sense of universal force,' the English artist and sculptor Barbara Hepworth said of stone as she fondled two dark pebbles, 'a silent image of our desire for survival, peace and security.'

Stone reverence is present and visible in the relics of Ancient Celts and Britons, where standing stones and dolmen were placed to mark sacred sites and an alignment with the heavens. The Mayans venerated stones and believed jade, in particular, held divine powers; stones held the spirits of their ancestors and were used in rituals or to indicate sites of sacred geography. In Judaism, there is a tradition for placing a stone on a grave to show that it has been visited and some believe that this tethers the soul to this realm. In Hinduism stones from sacred sites may be revered for the divine essence they contain. Meteoric stones are venerated in Native North American cultures. And Greek cults, such as that of Artemis, are believed to have worshipped meteorites representing their idols. Our ancestors revered chosen stones for their embodiment of the eternal, their perceived energetic channel between Heaven and Earth and as a touchstone upon which the devotions of a believer might catch the light of heaven.

# Darkness Was the Universe

## Part I

### *Conclusion*

It is 2013. It's a Friday. Miles outside of the city of Chelyabinsk, a man waits at a red light in west-central Russia on a freezing February morning; light is still winnowing into the blue-grey sky. Despite the clock on the dashcam reading 09:20:20, the Sun has not yet climbed above the horizon of dark forest that looms over the road and stretches for miles. Behind the forest are the Urals: a forbidding, isolated mountain range that runs for 1,600 miles (2,575km) – from the Arctic Ocean on Russia's coast to north-west Kazakhstan. It is a remote corner of the world and strange things happen around these parts.

Outside his car, there is snow on the ground and the first rays illuminate the gold dome of a monument on the man's right. He looks to the east, his direction of travel, where the Sun is slowly rising and the sky is blooming pink. But another sun appears from the blue of dawn: a white-green sun that burns like an otherworldly flare and falls as a blast of white. This is a 10,000-ton rock – a superbolide – entering the Earth's atmosphere and exploding at a height of around 18 miles (29km) above the ground, with the force of 30 atomic bombs.

CCTV footage of this moment shows office workers being shaken to the ground and cowering by their desks as tens of thousands of windows are smashed – the breaking of so much glass sounds like an upturned rainstick; papers fly into the air; doors are blown off in warehouses; and the sound above the city, like a giant fireworks display, reverberates through the air, making car alarms blare and dogs bark at the sky.

From a tower block, through a blown-out window frame, a man films the aftermath of the Chelyabinsk asteroid, now a towering plume of smoke, in disbelief. But drivers on the road below had continued on their way

– some driving directly towards the burning orb – and some people on the ground had continued walking along the street, barely even looking at the combusting sky. As one commenter on a video of the event writes, 'In Russia, we scare the crazy things.'

As the event occurred over the city of Chelyabinsk it caused significant disruption and damage, and footage of breaking windows and shaking buildings dominated on the news and the internet at the time. But the largest meteor fragment that landed on Earth's surface came down in a lake, missing the city, with its nuclear reactor and atomic waste storage facility, by around 25 miles (40km). Though many were injured as a result of its shockwaves, the human cost was relatively small compared to the size of the object.

Because the asteroid was coming from the direction of the Sun, telescopes were blinded and unable to detect it, and it doesn't take much to imagine the political, social and geographical – and later, cultural – upheaval such an event would have if it had hit a densely populated country, such as the UK. Some internet chat communities almost gleefully imagined it and an astrophysicist at Oxford University estimated that if the event had happened over London, people's hats would have been blown off in Edinburgh.

In 2018, a year after the asteroid 'Oumuamua' had caused a stir as it safely passed Earth, I spoke to Will

Sutherland, Senior Lecturer in Astronomy at the School of Physical and Chemical Sciences, Queen Mary University of London. We talked hypothetical asteroids and their potential for collisions. 'The asteroid would pop up first on the Astronomer's Telegram,' Sutherland told me, 'a service where people want quick follow-up observations – particularly for supernova, comets and asteroids. Most of it is world public, so anyone can read it. And most observations are quite dull: we observed such and such a star, with such and such a telescope. If this hypothetical asteroid popped up, they would follow up on it for the next few nights to see where it's going, and calculate an approximate orbit. And pretty quickly it would become evident if it's going to come close – they might not know initially if it will collide, but will know if it will come pretty damn close. Then there is a protocol to avoid panic. And we have something called the Torino Impact Scale, which is like the Richter scale but for asteroids, 0–10. There are a few asteroids we have detected that have numbers between 3 and 5 on the Torino Impact Scale, and there is one called Apophis, which is a little scary, because it's going to have a near-miss of the Earth in 2029.'

The Apophis asteroid is named after the ancient Egyptian god of chaos: on initial observation it was deemed there could be a small chance of impact, for which we have certain mitigation protocols, including nudging or exploding asteroids. But observations since confirm that Apophis will safely pass Earth on 13 April 2029, with NASA planning a mission to study the asteroid during this event. However, Apophis will return, and again there will be a chance of collision. As Sutherland explained, 'There is this thing called a "keyhole", where if the asteroid gets to *just* the right distance from the Earth it will get a gravitational kick that ensures it comes back and hits the Earth next time, in something like 2136. But most of the time, the kick isn't right.' So it floats through the nirvana of space, beautiful and harmless as a bubble for now.

As for the Chelyabinsk Event, there were no deaths, buildings were reconstructed, glass was replaced and damage to the ecosystem, and lasting damage to nearby towns and cities, was minimal thanks to the asteroid exploding around 18 miles (29km) above the Earth's surface, and its fragments landing where they did. The Chelyabinsk Event is one of several asteroid impacts the Earth has endured during human, and pre-human, history.

Black and white photographs from the aftermath of another, older event show a society of trees that look as if they have been cut down mid-stride as they tried to outrun their fate. The tall larches and conifers of the massive taiga forest lie charred and ruined above scorched grass

and freshly fallen snow. Not even the deepest root system would have stood a chance from a blast of 30 megatons.

Russia had been struck by what was most likely an asteroid, this time in rural Siberia. It was the largest asteroid in recorded history: the Tunguska Event of 1908. In Moscow State University you can read the story the trees had to tell in the rings of their trunks. It is one of total decimation. The asteroid impact destroyed an estimated 1,241 square miles (3,214 sq. km) of forest, and felled around 80 million trees. The shockwave was in the shape of a butterfly and, eerily, near the blast's epicentre, as in the eye of a storm, trees remained standing, calmly, but were stripped of bark and branches.

Though it is known as an impact event, as at Chelyabinsk, the Tunguska asteroid is thought to have exploded in the air, 3–6 miles (5–10km) above Earth, ensuring no impact crater was found. Because of the area's remote location, scientific investigation of the event was delayed. It was then further set back by the outbreak of World War I, which caused between 15 and 22 million more deaths than the greatest asteroid event in human history. As the taiga is so sparsely populated, there were only two unconfirmed reports of human casualties. But the devastation on the natural world remained enormous. Hundreds of reindeer were killed in the blast and they were a primary resource of food, warmth and transport for local tribes.

When I M Suslov, a Soviet ethnographer and geographer, travelled to the site in 1926, he spoke to a man called Chuchan from the Shanyagir tribe, who remembered the event well. He had been a child sleeping by the river next to his brother, Chekaren, when it had occurred. 'We heard whistling,' Chuchan told Suslov, 'and felt strong wind. Chekaren said, "Can you hear all those birds flying overhead?" We were both in the hut, couldn't see what was going on outside. Suddenly, I got shoved again, this time so hard I fell into the fire … There was noise beyond the hut, we could hear trees falling down … Then the thunder struck. This was the first thunder. The Earth began to move and rock, the wind hit our hut and knocked it over … Then I saw a wonder: trees were falling, the branches were on fire, it became mighty bright … as if there was a second sun, my eyes were hurting, I even closed them … And immediately there was a loud thunderclap. This was the second thunder … We looked at the fallen trees, watched the tree tops get snapped off, watched the fires … Now I remember well there was also one more thunder strike [the fifth], but it was small, and somewhere far away, where the Sun goes to sleep.'

The taiga forest supports tens of thousands of species – such as elk, brown bear, wolf, wolverines, eagle owls, capercaillie and Siberian swifts – and the impact the event had on these species would, undoubtedly, have been catastrophic. However, even by the time Suslov visited 14 years later, the forest had started to regenerate, and today it is a nature reserve, and life as we have come to know it goes on.

Rock is transformed into glass-like beads that rain down on an Earth before the concept of time. Sixty-six million years ago, an asteroid hit what is now the Yucatán Peninsula in Mexico. It was a carbonaceous chondrite asteroid that had travelled around 200 million miles (320 million km) from the icy reaches of the asteroid belt to hit Earth at 25,000mph (40,000km/h). While the Chelyabinsk asteroid was between 60 and 65ft (18–20m) wide, and the Tunguska asteroid between 165 and 265ft (50–80m), the Chicxulub asteroid was 6–9 miles (10–15km) across.

Its collision with Earth released energy over a billion times greater than the bomb that was dropped on Hiroshima in 1945 – humankind's current measure of destruction. The aftermath of the impact – due to the huge quantities of sulphur released into the air – resulted in the equivalent of a nuclear winter, known as an 'impact winter'. Temperatures plummeted by between 15 and 26°C. There was no sunlight and no moonlight. Acid rain and this hostile weather system heralded the long and dismal end of the non-avian dinosaurs, a species that had existed for 165 million years.

A few million years later, a now-famous literary group that included the likes of Mary Shelley and Lord Byron holed themselves up in a villa near Lake Geneva after the volcanic explosion of Mount Tambora in Indonesia in 1815. There, they endured the subsequent 'Year Without a Summer', where the eruption caused a volcanic winter in the spring. Endless palls and plumes of dust filled the atmosphere, drifted up into the stratosphere, and prevented sunlight from reaching the Earth – causing dimness all day, and global temperatures to plummet. In this environment, Mary Shelley summoned blackness into the human spirit writing *Frankenstein*, and Byron wrote the poem 'Darkness' about what he beheld: an echo of what had been witnessed through the eyes of a dinosaur 66 million years earlier.

*I had a dream, which was not all a dream.*
*The bright sun was extinguish'd, and the stars*
*Did wander darkling in the eternal space,*
*Rayless, and pathless, and the icy earth*
*Swung blind and blackening in the moonless air.*

Though pre-dating humankind, the Chicxulub asteroid is the most significant asteroid event in humankind's history – it changed the course of life on Earth. But beyond the extinction of a species that thereby removed the competition necessary for our mammalian ancestors to thrive, the influence of the asteroid within the socio-political or cultural realms of humanity might appear relatively minimal. However, the memory and the promise of another asteroid lowers like a dark cloud over our collective subconscious, an ultimate dread.

> *The clouds perish'd; Darkness had no need*
> *Of aid from them—She was the Universe.*

It is a dread we revel in, catastrophe is part of our DNA: it is how all life began. Perhaps it is why we like to stand out and watch the waves during a storm, why we get a kick out of watching sheet lightning at the end of summer, why we dream of supernovas and why we dare to make a wish on a burning piece of rock.

# Darkness Was the Universe

## Part II

### *Epilogue*

When there are no sparks in tailspin, when there is no distant comet charging towards the Sun, cruelly obscured by cloud at sundown, we look up and are blown away, out of ourselves, by the Moon. It is unfailingly breathtaking, gentle on the retina in a way the Sun never can be; it encourages our gaze, pulls it. So, we watch it rise, huge and yellow at the bottom of a drive as a harvest supermoon; or red and mythic as a winter worm moon above the sea; it glows white-blue as it sails through the sea of black and sinks behind the land. It is the luminous jewel above the diadem of our exosphere, and its surface is the last visible location that mankind has not populated, utilised, or colonised; for that reason in itself it serves as a primal relief. The Moon is the guardian of the lonely, companion of the hopeful, sidekick of the adventurer, the confidante of the lovers, and a witness to the grief-stricken. It has a dark side. It has a far side. And its benevolent fellowship depends upon all sides remaining beyond our reach.

The dark side of the Moon that has been popularised is really just the far side of the Moon, which, though not visible from Earth, receives its quota of sunlight. Over most of the lunar surface, sunlight illuminates its white crags and pours shadow into its craters and basins, its regions of volcanoes and apparent seas of tranquility, helping to create the Moon's distinctive face. But there is darkness on the Moon, there are places where sunlight does not reach, a true dark side, and that is located at the Moon's south pole.

The Moon's south pole is a dusty plane of darkness and shadow. Within it lies the South Pole-Aitkin Basin, inside which is a trauma: one of the largest impact craters in the Solar System.

The Moon features many craters from the early bombardment period, when our Solar System was host to migrating swarms of blind rock, dust and ice, but this 1,600-mile (2,575km) crater was created by an asteroid estimated to have been between 100–160 miles (160–260km) in diameter, travelling from the main asteroid belt. That collision between the asteroid and the Moon occurred 4.3 billion years ago – a mere few hundred million years after the Moon was formed and was still cooling from its beginning as a spluttering laval mass. It is thought that a body detected under that basin on the lunar south pole may be a remnant of this asteroid. But we don't know for sure. The South Pole-Aitkin Basin (SPA) has secrets about its past and we don't like things keeping secrets from us.

At the time of writing, the NASA-led mission Artemis III is due to launch in 2028. This will take humankind to the surface of the Moon for the first time in over 50 years. Astronauts will visit the true dark side of the Moon, the lunar south pole, for the first time. The hope is that the study of this asteroid impact crater will increase our understanding of the Moon's formation, gather samples from the Moon's interior, and perhaps even lead to the discovery of microbial life in moon ice – the frozen lunar water known to sit in permanent shadow.

But we have other motives beyond the noble cause of understanding driving us back to the Moon. As NASA writes on its website, 'This mission will usher in a future in which humans consistently access the Moon, and human planetary exploration missions are within reach.' There will be 'infrastructure deployment in the lunar south pole region, focusing on understanding the Moon's composition, history, and resource potential'. All in the aid of the 'long-term goals of establishing a sustained lunar presence'.

It may be that one day, meteors, comets and asteroids are the only celestial objects visible to the naked eye that we have not populated permanently in some way. While there will be many who rejoice at the prospect of human habitation on the Moon, with the installation of infrastructure on its surface and within its orbit, there will be others who question whether, after conducting our scientific enquiry, we should leave no further trace.

That we are again considering the impact of our behaviour in the heavens returns us to a quasi-medieval mindset. Only now we can travel into that realm of perceived divinity where a meteor, comet or asteroid, such as the one that hit the Moon, were once seen as the machinery of supreme beings. We do not need to look up: we can look down and ask the hand of God to take us to its beating heart, to lead us deeper into the planes of angels, further into the cold black lands of wandering souls, for Mars lies at the outermost periphery of the asteroid belt.

Though we now know exactly what hit the Moon (down to its probable molecular structure), we know exactly what that comet that comes every 75 years is, exactly what the tears of St Lawrence are every August, we know it is ice and dust, we know it is rock, but we also cannot unknow that it is a dragon, a blessing, a mourning, an omen, a wish, because of everything, everyone, that has come before us. To regard a shooting star is to see a raven in the Sun: it is to be both dazzled and haunted. The image of a shooting star is like the representations of the shamanic cosmos, it is hardwired into our neural pathways. At one time that image was a signpost. In some interpretations, it still is.

These hanging, these broom, these pheasant, these guest, these free, these uncommon stars will continue to be the symbolic keys to our most ancient ancestors and the origins of our cosmos, as well as the keys to our many potential futures. More than anything else, or, perhaps, at the core of everything else, they urge us to accept that although we may one day have dominion over the heavens, we are not gods. And what a relief that is. We will never know what is going to happen next, but it might be something wonderful.

# Comets' Return

- The comet of 44 BCE, known as Caesar's Comet, is considered a non-periodic comet unlikely ever to return having travelled into deep space.
- The comet that was observed in 678 CE and identified as the first comet recorded in British history is now identified as Comet C/2025 A6 (Lemmon). It last returned to the inner solar system in 2025 and is expected to return sometime between 3175 or 3421.
- The twin comets observed in 729 AD are not expected to return until approximately 3421.
- The Great Comet of 905 (designated C/905 K1) has a long, parabolic orbit and may not return to the inner solar system for tens of thousands of years.

- The Comet of 1491 (designated 1491 B1 or C/1490 Y1) is believed to have disintegrated shortly after its appearance in the skies above Earth.

- The exact identity of the Great Comet of 1532 is unclear, it has previously been identified as C/1532 R1 but many believe this is now incorrect. It may have been the Comet 153P/Ikeya-Zhang, discovered in 2002 and expected to return to the inner solar system in around 2360.

- The Great Comet of 1577 is not a short-period comet and not known when, if ever, it will return.

- The Great Comet of 1811 (C/1811 F1) is a long-period comet and is not expected to return to the inner Solar System until around 4800.

- Donati's Comet, last observed in 1858, is expected to return sometime between 3600 to 3759.

- The Comet Thatcher, last observed in 1861, is expected to return to the inner solar system in 2283. It is long-period comet with an orbital period of approximately 415–422 years, and the debris from this comet forms the Lyrid meteor shower.

- The sun-grazing Great Comet of 1882 (C/1882 R1) broke apart into between four to six fragments which have long, elliptical orbits. They are expected to return to the inner solar system between approximately 2487 and 2719

- Halley's Comet (last observed in 1986) will next return in 2061.
- The comet Tsuchinshan-ATLAS, last observed in the inner solar system in 2024, will return in 82024.

# Dark Sky Zones

It is worth noting that for optimal viewing in all 'Dark Zones' stargazers should aim to visit during a new moon for minimal natural light interference.

## UK

### *Exmoor – England*

Located in Devon, in the South West of the UK, Exmoor the first place in Europe to be designated an official Dark Sky Zone in 2011. With its minimal light pollution, expansive horizons and often a quick sea breeze clearing the cloud, there are thousands if not millions of stars visible to the naked eye, including the Milky Way. Having lived there, I can attest it is one of the best places I have been to view the night sky.

### *Elan Valley – Wales*

Located in mid-Wales, Elan Valley Dark Sky Park achieved its status in 2015. With its relatively remote location, surrounded by the Cambrian Mountains, lakes and a nature reserve, resulting in minimal light pollution, I can highly recommend it as an incredibly picturesque location to watch a meteor shower.

### *The Isle of Rúm – Scotland*

The Isle of Rúm, located off the west coast of Scotland, was officially designated as Scotland's first Dark Sky Sanctuary in August 2024. I spent time just a little further south, on the Sound of Mull recently and when we finally got a clear, freezing night I can confirm when I looked up at the black I saw more stars than I have ever seen in my entire life. The area is considered a premier destination for star-gazing.

### *The Kerry International Dark-Sky Reserve – Ireland*

Spanning 700 $km^2$ and located along the west, Skellig Coast of Ireland, on the Iveragh Peninsula this dark sky reserve has been recognized for its exceptionally low light pollution. It boasts incredibly naked-eye views of the Milky Way and even nebulae, ensuring it is regarded as one of the best stargazing location sin the Northern Hemisphere. It is one of the top stargazing spots in the Northern Hemisphere.

## UNITED STATES

Central Idaho Dark Sky Reserve, located in Idaho, is regarded as the nation's only gold-tier (darkest) reserve, meaning it will typically have optimum conditions for astro-tourists.

Death Valley National Park, located in California, is known for having almost-guaranteed clear night skies, with over 300 clear days annually and for being removed from urban light.

Oregon Outback International Dark Sky Sanctuary covers around 2.5 million acres (roughly 3,900 square miles) and is one of the largest Dark Sky Sanctuary's in the world.

Glacier National Park, located in Montana, is renowned for its unpolluted skies, offering optimal locations for stargazing and for viewing the Milky Way. It was designated in 2017 and, combined with Canada's Waterton Lakes National Park, it was the world's first international trans-boundary Dark Sky Park.

Mauna Kea, Hawaii, located on a dormant volcano, is one of the best dark sky locations in the world. The conditions of low humidity and minimal light pollutiontion and high altitude combine to make exceptional stargazing. It is also home to telescopes at the summit.

## EUROPE

### *France*

Pic du Midi (Pyrenees) was designated 2013 and was the first Dark Sky Reserve in continental Europe, gaining its status in 2013, and covers 3,112 $km^2$ featuring a completely unlit 620 $km^2$ core zone and an observatory.

### *Portugal*

Portugal boasts the Dark Sky Alqueva reserve in the Alentejo region, which was the world's first site to be certified as a 'Starlight Tourism Destination' by the Starlight Foundation and supported by UNESCO. Covering over 3,000 square kilometres around Lake Alqueva, it offers exceptional, low-light-pollution conditions for stargazing

### *Spain*

The Sierra Morena Starlight Reserve in Andalucía is the world's largest starlight destination, spanning over 400,000 hectares and 57 municipalities across the northern provinces of Huelva, Seville, Córdoba, and Jaén.

### *Germany*

Westhavelland Nature Park (Brandenburg), known as the first designated dark sky reserve in Germany, is located about 70 km west of Berlin, offering exceptionally dark skies.

### *Poland*

Izera Dark Sky Park (Izerski Park Ciemnego Nieba) was established in 2009 and was the first dark-sky park in Poland, and the world's first transboundary dark-sky park (located on the Polish-Czech border). It is located in the southwest of Poland in the Jizera Mountains.

## AFRICA

### *St Helena*

Before Edmond Halley correctly identified the comet now of his namesake's returning orbit from his observatory in Islington, London, he travelled to St Helena, a remote island in the South Atlantic, 2,000 km from the West coast of Africa. There – with its view of the stars in both the Northern and Southern Hemispheres – he established an observatory in 1677 with the aim of a comprehensive, telescopically determined star catalogue.

His observatory has since been demolished after falling into disrepair, but there is a new observatory, called Halley's Mount Observatory which is located on the fringes of St Helena's Peaks National Park that the public are able to visit.

At present, the island is actively working to become an official International Dark Sky Place, with legislation having been passed to protect its exceptionally dark, low-light-pollution skies.

### *Namibia*

NamibRand Nature Reserve. This was Africa's first International Dark Sky Reserve and it offers stargazers some of the darkest, and least light-polluted skies on Earth.

### *South Africa / Botswana*

!Ae!Hai Kalahari Heritage Park is a certified International Dark Sky Sanctuary, boasting incredible viewing conditions.

Sutherland and The Karoo is home to the Southern African Large Telescope (SALT). This region is renowned for its 80% cloud-free nights and near-zero light pollution.

## SOUTH AMERICA

### *Chile*

The Atacama Desert, located to the north of the country, is considered to be one of the best dark sky zones on Earth, with many locations offering professional telescope tours of their night skies, which are typically clear for 300 nights a year. Popular destinations include San Pedro de Atacama and the Elqui Valley, which was the world's first International Dark Sky Sanctuary).

### *Argentina*

El Leoncito National Park, located in the San Juan province, is considered one of the best stargazing locations in South America and offers exceptionally clear, low-light-pollution skies. It is also home to two major observatories: the Complejo Astronómico El Leoncito (CASLEO) and the Dr Carlos U Cesco Observatory.

### *Bolivia*

The vast, high-altitude salt flats of Salar de Uyuni provide panoramic, unobstructed views of the night sky with near-zero light pollution. Its high-altitude (over 3,500m), dry climate, and remote location ensure conditions are optimal for viewing the Milky Way and southern constellations.

### *Colombia*

The Tatacoa Desert is a starlit reserve that is considered an emerging, optimal offering dedicated astronomical observatories, such as Astrosur and OATA, which provide telescope viewings, guided tours, and also host an annual astronomy festival in mid-August.

## INDIA

The Hanle Dark Sky Reserve (HDSR) in Ladakh offers viewers exceptional, zero-light-pollution views of the Milky Way at over 4,500 meters.

Spiti Valley, Himachal Pradesh is another high-altitude location that is high-regarded highly-regarded for its remote and clear skies, ideal for astrophotography.

## AUSTRALIA

Warrumbungle National Park, located in New South Wales, was the first officially designated International Dark Sky Park and offers exceptionally low light pollution for stargazers.

Arkaroola Wilderness Sanctuary, located in South Australia, became only the second certified Dark Sky Sanctuary in Australia. Different from Dark-Sky Parks or Reserves, Sanctuary's are often more remote and require special considerations for approval, such as a proven commitment to conservation, preservation and raising awareness.

River Murray Dark Sky Reserve, located in South Australia, pans over 3,200km$^2$ and stretches along 80km of the Murray River. It was accredited in 2019 and features some of the darkest skies in the world, with Swan Reach Conservation Park at its heart.

## CHINA

Lenghu Dark Sky Park is located in the Gobi desert. This area is rapidly becoming a premier destination for astronomical observation due to its high altitude and strict lighting regulations.

Huanglong International Dark Sky Park, located in Sichuan, was the first certified International Dark Sky Park in mainland China. It offers pristine, high-altitude (3,900-4,000m) viewing.

## JAPAN

Bisei Town, Okayama – as mentioned in this book, this town is known as the 'Town of Beautiful Stars'. It is an area that pioneered light pollution prevention in Japan and the Bisei Astronomical Observatory is a key attraction. You can visit during the Tanabata Festival on the seventh day of the seventh month to celebrate the union of the Shokujo (also known as Orihime), the celestial weaver of Japanese folklore and her spouse.

Kozushima Island is regarded as a 'Dark Sky Island' and is located 180 km south of Tokyo.

Iriomote-Ishigaki National Park, located in Okinawa, was Japan's first International Dark Sky Park, known for subtropical, pristine night skies.

# Acknowledgements

I often think how a writer is less a product of what they want to write and more a product of what they get published, and so it is with enormous gratitude that I thank my editor, Rebecca Armstrong, for giving me the opportunity to write this book – on a subject that has been a fascination, a fixation, of mine for my entire life. Thank you to Joe McLaren for bringing the book to life with such beautiful illustrations. And a huge thank you to the entire team at brilliant Batsford Books – to James Kellow, to Eoghan O'Brien, to Mia Autumn Roe, to Katie Hewett, and the gimlet-eyed Ian Allan. I am very proud to be published under Batsford and constantly inspired by your vision.

Thank you, as always, to my brilliant agent and friend, Max Edwards, without whom none of this would have been possible.

Thank you to Will Sutherland, Senior Lecturer in Astronomy at the School of Physical and Chemical Sciences, Queen Mary University of London, for talking to me back in 2018 about hypothetical asteroids; and to all the incredible researchers, anthropologists, metallurgists, astrophysicists, meteoriticists, astronomers, astrologers, alchemists, historians, art critics, musicologists, and folklorists whose work informed and inspired all of these essays.

Thank you to *When The Sky Falls'* early readers for taking time out of your busy days.

Thank you to my friends and family who seem to love me even when I am wild-eyed and unwashed, citing some new fascinating bit of research I just *have* to add. And it feels rude not to thank the sky and the bodies that travel through it, for being so unrelentingly beautiful and awe-inspiring.

For Loki, who flies faster and shines brighter than any meteor.

# About the Author

Jade Angeles Fitton is an author, journalist and poet with a special interest in nature writing, cultural history and folklore. Her critically acclaimed debut, *Hermit: A Memoir of Finding Freedom in a Wild Place* (Hutchinson Heinemann), was published in 2023, and her other writing has appeared in the *Guardian, The Times, The Spectator, Vogue, The New Statesman* and many more. Her poetry has been highly commended in The National Poetry competition and shortlisted for The Moth Poetry Prize. Her short fiction has been broadcast on BBC Radio 4 and published in the likes of *The London Magazine.* She lives in rural Devon.

# References

## Introduction

Matthieu Gounelle, 'The meteorite fall at L'Aigle and the Biot report: exploring the cradle of meteoritics', Geological Society, London, Special Publications, 2006

## Part One

### *Melancholia*

Mark Hutchinson, 'The Art of Melancholy', *Times Literary Supplement*, 20 December 2005

Brian Boucher, 'Art Bites: The Mystery Object in a Famous Dürer Print', Artnet.com, February 2020

Peter Lynch, 'The enduring enigmas of Albrecht Dürer's "Melencolia I"', The *Irish Times*, November 2015

The Painting of Melancholia, https://paintingmelancholia.blogspot.com/2010/08/albrecht-durer-british-museum-london.html

Fredrika Thelandersson, Chapter 2, 'A Historical Lineage of Sad and Mad Women', *21st Century Media and Female Mental Health*, Palgrave Macmillan, 2022

Cecilia Tasca, Mariangela Rapetti, Mauro Giovanni Carta and Bianca Fadda, 'Women and hysteria in the history of mental health', *Clinical Practice and Epidemiology in Mental Health*, Vol. 8, 2012

## *The School of Night*

L McCormick-Goodhart, 'Shakespeare and the stars', *Popular Astronomy*, Vol. 53, 1945

The Marlowe Society, www.marlowe-society.org

## *The Missing Piece*

Deborah Wye, *Louise Bourgeois: The Complete Prints & Books*, MoMA, 2012

## *White Star*

Eyewitness account of the sinking of the *Titanic*, 1912, Gilder Lehrman Institute of American History https://www.gilderlehrman.org/sites/default/files/inline-pdfs/07640_FPS.pdf

*Titanic* – Deleted Scene – Shooting Star [HD], https://www.youtube.com/watch?v=DNIx3vpKVc8

## *The Messenger*

Stillman Drake (abridged, translation), *The Assayer (1623) by Galileo Galilei*, Stanford University, 1957

C J Humphreys, 'The Star of Bethlehem – a Comet in 5 BC – and the Date of the Birth of Christ', *Quarterly Journal of the Royal Astronomical Society*, Vol. 32, No. 4, 1991

Bartolomé Esteban Murillo, *The Adoration of the Magi*, Toledo Museum of Art, https://emuseum.toledomuseum.org/objects/55299/the-adoration-of-the-magi

Giotto, *The Adoration of the Magi*, Art and the Bible, https://www.artbible.info/art/large/260.html

## *Christmas Eve*

Nikolai Gogol, *Christmas* Eve, https://musicwebinternational.com/2024/02/rimsky-korsakov-christmas-eve-naxos

Richard Taruskin, '"Kitezh": Religious Art of an Atheist', *The New York Times*, February 1995

## *Pines Where Stars Have Fallen*

M Shima, S Murayama, H Yabuki, and A Okada, 'Petrography, mineralogy and chemical composition on the chondrite Nogata, Nogata-shi, Fukuoka-ken, Japan, oldest Observed fall in the world', *Abstracts of Papers Presented at the Forty-Third Annual Meeting of the Meteoritical Society*, 1980

Mitch Noda and Jesse Piper, 'Nihon Inseki or Japanese Meteorites', *Meteorite Times Magazine*, 2023

## *The Brightest Point*

Robert G Aitken, 'Edmund Halley and Stellar Proper Motions', *Astronomical Society of the Pacific Leaflets*, Vol. 4. No. 164, 1942

'Newton's observations of the comet of 1682 (MS Add.9597/2/17)', University of Cambridge Digital Library, https://cudl.lib.cam.ac.uk/view/MS-ADD-09597-00002-00017/1

Royal Museums Greenwich, 'Edmond Halley, 1656–1742, Astronomer Royal', https://www.rmg.co.uk/collections/objects/rmgc-object-14208

## *Comet Vintages*

Thomas Hardy, 'The Comet at Yell'ham', *Poems of the Past and the Present*, Harper & Brothers, 1902

James W Watts, *Donati's Comet, 10 October 1858*, The Royal Society Print Shop

'Comet Vintages', *Wikipedia: The Free Encyclopedia*, Wikimedia Foundation, 7 September 2025, https://en.wikipedia.org/wiki/Comet_vintages

### *Love's Arrow*

Britannica, 'Orpheus: Greek Mythology', https://www.britannica.com/topic/Orpheus-Greek-mythology

NASA, 'C/1861 G1 (Thatcher)', https://science.nasa.gov/solar-system/comets/c-1861-g1-thatcher

## Part Two

### *Cosmic Energy*

'New approach to "cosmic magnet" manufacturing could reduce reliance on rare earths in low-carbon technologies', Energy Interdisciplinary Research Centre, University of Cambridge, https://www.energy.cam.ac.uk/blog/research-highlight-new-approach-cosmic-magnet-manufacturing-could-reduce-reliance-rare-earth

Michel Prévot and David Dunlop, 'Louis Néel: forty years of magnetism', *Physics of the Earth and Planetary Interiors*, 2001

### *Buried Blades*

Dr Diane Johnson, 'Iron from the sky: Meteors, meteorites and ancient culture', OpenLearn, The Open University, 2023

### *Prescribing Stars*

Godfrid Storms, *Anglo-Saxon Magic*, Martinus Nijhoff, 1948

Lynn Thorndike, *The Place of Magic in the Intellectual History of Europe*, The Macmillan Company, 1905

James George Frazer, *The Golden Bough*, Macmillan and Co., 1890

## *The Iron Mountain*

Martin Appelt, Jens Fog Jensen, Mikkel Myrup, Henning Haack, Mikkel Sørensen and Michelle Taube, *The Cultural History of the Innaanganeq/Cape York Meteorite*, The Greenland National Museum and Archive, 2015

Emily Johnson, 'The Cape York Meteorite: Making an Impact on Greenland', The Henry M Jackson School of International Studies, 2019

'Fragments of Cape York', American Museum of Natural History, https://www.amnh.org/exhibitions/permanent/meteorites/meteorites/fragments-of-cape-york

## *Dark Markets*

Werner Menges, 'Meteorite Man Granted Bail', *The Namibian* (accessed via AllAfrica.com), September 2002

Richard Hamilton, 'Meteorite smugglers anger scientists', news.bbc.co.uk, April 2007

## *The Šaman's Amulet*

'Archaeologists discovered a meteorite fragment in a 9 thousand year old hut', Science in Poland, 2014, https://scienceinpoland.pl/en/news/news,400865,archaeologists-discovered-a-meteorite-fragment-in-a-9-thousand-years-old-hut

'Evenki Shamanism', Museum of Archaeology & Ethnology, Simon Fraser University, https://www.sfu.ca/archaeology/museum/exhibits/virtual-exhibits/russian-traditional-village-life/evenki-shamanism.html

Christine S VanPool, 'Shaman Pots, Sympathetic Magic, and Spinning Souls among the Medio Period Casas Grandes: Altered States of Consciousness in Other-than-Human Persons', *Religions*, 2024

### *The Winter Years*

Larry R Nittle, 'Presolar stardust in meteorites: Recent advances and scientific frontiers', *Earth and Planetary Science Letters*, Vol. 209, 2003

A N Nguyen, L P Keller, D E Brownlee and D J Joswiak, 'Mineralogy of Presolar Grains of Varied Stellar Origins in Giant Cluster IDP U2-20GCA', 54th Lunar and Planetary Science Conference, March 2023

Kate Golembiewski, 'Studying Galactic Star Formation with Presolar Grains', Robert A Pritzker Center, 2020

## Part Three

### *What Light is That?*

Christopher Marlowe, *Edward II*, Act V, Scene V, 1592

Haywood Williams Guion, *The Comet: or, the Earth, in Her Varied Phases, Past, Present and Future, as Deduced from the Highest and Most Reliable Authoritie*, E J Hale & Son, New York, 1869

### *A Bull with Heaven on His Back*

James R Murie, 'Ceremonies of the Pawnee, Part I: The Skiri', *Smithsonian Contributions to Anthropology*, 1981

Von Del Chamberlain, *When Stars Came Down to Earth: Cosmology of the Skidi Pawnee Indians of North America*, Ballena Press, 1982

### *Free Stars*

Henry Adams Bellows (translation), *The Poetic Edda* (1936)

### *Akkadian Omens*

Judith Kingston Bjorkman, 'Meteors and Meteorites in the Ancient Near East', *Meteoritics*, Vol. 8, No. 2, 1973

### *She Who Works Her Will*

A W Mair (translation), *Callimachus and Lycophron*, William Heinemann, London; G P Putnam's Sons, New York, 1921

### *Zmej*

Tsimafei Avilin, 'Meteor Beliefs Project: Belarussian meteor folk-beliefs', *WGN*, Journal of the International Meteor Organization, Vol. 34, No. 4, 2006

Roger Deakin, *Wildwood: A Journey Through Trees*, Penguin Books, London, 2008

### *Hanging Stars*

E G Mardon, A A Mardon and J Williams, 'The Eleven Observations of Comets 678AD and 1114AD Recorded in the Anglo-Saxon Chronicles', *Asteroids, Comets, Meteors*, 1991

Andrei Dorian Gheorghe, Roy Watson and Alastair McBeath, 'Meteor Beliefs Project: Birth and death superstitions associated with meteors in Romanian and British folklore', *WGN*, Journal of the International Meteor Organization, Vol. 34, No. 5, 2006

Umberto Dall'Olmo, 'Meteors, Meteor Showers and Meteorites in the Middle Ages: From European Medieval Sources', *Journal for the History of Astronomy*, Vol. 9, Issue 2, 1978

J A Giles, 'William of Malmesbury's Chronicle of the Kings of England. From the earliest period to the reign of King Stephen', Bohn's Antiquarian Library, 1847

### *Everywhen*

Philip A Clark, 'An overview of Australian Aboriginal ethnoastronomy', *Archaeoastronomy*, Vol. 21, 2009

http://www.aboriginalastronomy.com.au/content/topics/craters

### *Astral Migration*

B Fraser, 'Presidential Address: Astronomy in the lives of the indigenous people of Southern Africa (Presidential Address)', *Monthly Notes of the Astronomical Society of Southern Africa*, 1995

### *The Flight of Souls*

Galina Ershova, 'Comets and meteors in the beliefs of ancient mayas', *Astronomical & Astrophysical Transactions*, Vol. 20, Issue 6, 2001

### *Water Chestnut*

K Yau, P Weissman and D Yeomans, 'Meteorite falls in China and some related human casualty events', Meteoritics, Vol. 29, No. 6, 1994

## Part Four

### *The Welkin Aflame*

E G Mardon, A A Mardon and J Williams, 'The Eleven Observations of Comets 678AD and 1114AD Recorded in the Anglo-Saxon Chronicles', *Asteroids, Comets, Meteors*, 1991

A A Mardon and E G Mardon, 'The Venerable Bede's Observation of the Comet of 729 AD', 65th Annual Meteoritical Society Meeting, 2002

### *A Smoke Banner*

R C Kapoor, 'Tales from India. 1: Meteor showers in Classical and Colonial Sources', *Journal of Astronomical History and Heritage*, Vol. 25, No. 3, 2022

Patrick Das Gupta, 'Comets in Indian Scriptures', presented at the 8th conference on The Inspiration of Astronomical Phenomena, 2013

### *Planta Genista*

Geoffrey of Monmouth, *History of the Kings*, The National Library of Wales, c.12th century, https://www.library.wales/discover-learn/digital-exhibitions/manuscripts/the-middle-ages/history-of-the-kings

Christopher Michael Berard, *Arthurianism in Early Plantagenet England: From Henry II to Edward I*, Boydell & Brewer, 2019

Karen Cherewatuk, review of the above title, https://www.journals.uchicago.edu/doi/full/10.1086/727239

### *House of Fire*

G Latura, 'Caesar's Comet and the Birth of the Roman Empire', American Astronomical Society, 2021

Silver coin, head of Augustus wearing oak-wreath (obverse), comet with eight rays and tail upwards (reverse), https://www.britishmuseum.org/collection/object/C_R-6068

Keith T Smith, Martin Ward, 'Views: Caesar's Comet', *Astronomy & Geophysics*, Vol. 55, Issue 5, 2014

### *Who Can Tell Good From Bad?*

D W Pankenier, 'Heaven-Sent: Understanding Cosmic Disaster in Chinese Myth and History', *BAR International Series*, 1998

Claude Larre, Isabelle Robinet and Elisabeth Rochat de la Vallée, 'The old man lost his horse', *Les grands traités du Huainan zi*, 1993

'Emperor Xuanzong of Tang', *Wikipedia: The Free Encyclopaedia*, Wikimedia Foundation, 27 November 2025, https://en.wikipedia.org/wiki/Emperor_Xuanzong_of_Tang

### *Uncommon Star*

Benjamin Woolley, *The Queen's Conjuror: The Life and Magic of Dr Dee*, Flamingo, 2002

Mary Esther Trueblood, 'John Dee and his "Fruitful Preface"', *Popular Science Monthly*, Vol. 77, 1910

Henry Howard, 'A defensatiue against the poyson of supposed prophesies not hitherto confuted by the penne of any man, which being grounded, eyther vppon the warrant and authority of olde paynted bookes, expositions of dreames, oracles, reuelations, inuocations of damned spirites, iudicialles of astrologie, or any other kinde of pretended knowledge whatsoeuer, de futuris contingentibus: haue beene causes of great disorder in the common wealth, cheefly among the simple and vnlearned people: very needefull to be published at this time, considering the late offence which grew by most palpable and grosse errours in astrology.' In the digital collection Early English Books Online 2. https://name.umdl.umich.edu/A03738.0001.001. University of Michigan Library Digital Collections. Accessed July 3, 2025.

### *Black Stone*

Elsebeth Thomsen, *'New Light on the Origin of the Holy Black Stone of the Ka'ba'*, Meteoritics, Vol. 15. No. 1, 1980

Kathryn Weedman Arthur, Ran Barkai, Catherine Allen, Ella Assaf Shpayer, Bar Efrati, Meir Finkel, Dov Ganchrow, Rachel A Horowitz, Vlad Litov, Marlize Lombard, Paul Sillitoe and Edward Swenson, 'Ancestral Stones and Stone Stories: Reimagining Human Relationships with Stone from the Paleolithic to the Present', *Journal of the World Archaeological Congress*, Vol. 20, No. 1, Springer, 2024

## *Darkness Was the Universe, Part I*

I M Suslov (recorded), 'Testimony of Chuchan of the Shanyagir tribe', 1926

'Meteor Hits Russia Feb 15, 2013 – Event Archive', https://www.youtube.com/watch?v=dpmXyJrs7iU

'The Chelyabinsk meteor', The Planetary Society, https://www.planetary.org/video/the-chelyabinsk-meteor

'Chicxulub crater', *Wikipedia: The Free Encyclopedia*, Wikimedia Foundation, 5 December 2025, https://en.wikipedia.org/wiki/Chicxulub_crater

Lord Byron, *'Darkness'*, The Poetry Foundation, https://www.poetryfoundation.org/poems/43825/darkness-56d222aeeee1b

## *Darkness Was the Universe, Part II*

'Artemis III', NASA, https://www.nasa.gov/mission/artemis-iii

# Index

G



H



I

J



K



L



M